Test Item File for

Trigonometry

SIXTH EDITION

Larson/Hostetler

HOUGHTON MIFFLIN COMPANY **Boston New York**

Publisher: Jack Shira
Managing Editor: Cathy Cantin
Development Manager: Maureen Ross
Development Editor: Laura Wheel
Assistant Editor: James Cohen
Supervising Editor: Karen Carter
Senior Project Editor: Patty Bergin
Editorial Assistant: Allison Seymour
Production Technology Supervisor: Gary Crespo
Executive Marketing Manager: Michael Busnach
Senior Marketing Manager: Danielle Potvin
Marketing Associate: Nicole Mollica
Senior Manufacturing Coordinator: Jane Spelman

Printed in the United States of America

ISBN 0-618-31800-3

1 2 3 4 5 6 7 8 9 – EB – 07 06 05 04 03

GENERAL TABLE OF CONTENTS

Chapter P Prerequisites

Section P.1: Review of Real Numbers and Their Properties

Objective 1: Represent and classify real numbers

1. Identify which of the following numbers are rational numbers.
$5, -\frac{3}{20}, -3, 10\pi, 15, 0.4, \sqrt{7}, -\frac{3}{25}$

(A) $-\frac{3}{20}, 0.4, -\frac{3}{25}, 5, -3, 15$ (B) $5, \sqrt{7}$ (C) $10\pi, 15$ (D) $\sqrt{7}, 15, -\frac{3}{25}, 5$

2. Write the rational number as the ratio of two integers.
$0.\overline{78}$

(A) $\frac{39}{5}$ (B) $\frac{39}{50}$ (C) $\frac{26}{33}$ (D) $\frac{39}{25}$

3. Identify the numbers that are *not* irrational numbers.
$25, \sqrt{5}, 3.56, 0, -7.3333, \frac{3}{5}, -8$

4. Use a calculator to find the decimal form of the rational number. If it is a nonterminating decimal, write the repeating pattern.
$\frac{7}{11}$

Objective 2: Order real numbers and use inequalities

5. Use inequality notation to describe the set.
It takes no fewer than four years to complete law school.

(A) $l < 4$ (B) $l \geq 4$ (C) $l \leq 4$ (D) $l > 4$

6. Identify the verbal description of the interval.
$[-10, -9)$

(A) All real numbers greater than –10 and less than or equal to –9

(B) All real numbers greater than –10 and less than –9

(C) All real numbers greater than or equal to –10 and less than –9

(D) All real numbers greater than or equal to –10 and less than or equal to –9

7. Place the correct symbol (<, >, or =) between the pair of real numbers.

$-\frac{7}{16} \bigcirc -\frac{2}{7}$

8. Verbally describe the subset of real numbers represented by the inequality. Then sketch the subset on the real number line. State whether the interval is bounded or unbounded.

$x > -5$

Objective 3: Find the absolute values of real numbers and find the distance between two real numbers

9. Jon is on a mountain 11,058 feet above sea level. Jeana is in a submarine 3850 feet below sea level. Which of the following can be used to find the difference between Jon's elevation and Jeana's elevation?

(A) $|-3850|-|11,058|$ (B) $|11,058-3850|$

(C) $|3850-11,058|$ (D) $|11,058-(-3850)|$

10. Identify the expression that corresponds to the distance shown by the bar.

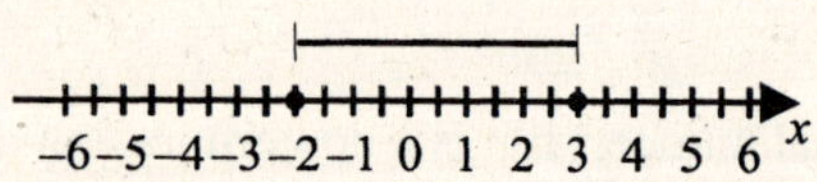

(A) $-2+3$ (B) $-2-3$ (C) $|-2-3|$ (D) $|2+3|$

11. Order the values from smallest to largest.

$|-16|, 9, |-7|, 2$

12. Evaluate the expression.

$|-3| - |-9|$

Objective 4: Evaluate algebraic expressions

Evaluate the expression for the given value of x.

13. $(4 \cdot x)\left(\frac{2}{3}\right)$ for $x = -3$ (A) $-\frac{1}{8}$ (B) $-\frac{8}{9}$ (C) -8 (D) $\frac{8}{3}$

Evaluate the expression for the given value of x.

14. $\dfrac{x^2-1}{1-x^3}$ for $x = 15$ (A) $-\dfrac{16}{241}$ (B) $-\dfrac{4}{49}$ (C) $\dfrac{1}{14}$ (D) None of these

15. $2x^2 - 9x + 3$ for $x = 2$

16. $3x^2 + 5x - 4$ for $x = -6$

Objective 5: Use the basic rules and properties of algebra

17. Identify which of the following is an example of the Associative Property of Multiplication.

(A) $7 \cdot 8 = 8 \cdot 7$ (B) $5 \cdot (7+8) = 5 \cdot 7 + 5 \cdot 8$ (C) $7 + 8 = 8 + 7$ (D) $7 \cdot (8 \cdot 5) = (7 \cdot 8) \cdot 5$

18. Identify the rule of algebra illustrated by the equation.
$99 + 0 = 99$

(A) Distributive Property (B) Additive Identity Property

(C) Associative Property of Addition (D) Commutative Property of Addition

19. Give an example of the Additive Identity Property.

20. Perform the operation. (Write the fractional answer in simplest form.)
$\dfrac{3}{5} \div \dfrac{5}{9}$

Section P.2: Solving Equations

Objective 1: Identify different types of equations

21. Which of the following is an identity?

(A) $2(x-2) = 4 - 2x$ (B) $2(x-2) = x - 4$

(C) $2(x-2) = 2x - 4$ (D) $2(x-2) = 2x - 2$

22. Which of the following is a conditional equation?

(A) $\dfrac{x}{4} \div \dfrac{x}{5} = \dfrac{5}{4}$ (B) $\dfrac{4}{x} \cdot \left(-\dfrac{5}{x}\right) = -\dfrac{20}{x^2}$ (C) $\dfrac{4}{x} - \dfrac{5}{x} = -1$ (D) $\dfrac{4}{x} - \dfrac{5}{x} = -\dfrac{1}{x}$

23. Determine whether the equation is an identity or a conditional equation.
$4(x+3)=8x-4x+12$

24. Determine whether the equation is a conditional or an identity.
$x^2+8x+20=(x+4)^2+4$

Objective 2: Solve linear equations in one variable and equations that lead to linear equations

Solve the equation and check your solution.

25. $4=4(x-2)+2-3x$ (A) 8 (B) 14 (C) 4 (D) 10

26. $3(x+4)+3=4(x+5)+4$ (A) -2 (B) 10 (C) -9 (D) 33

27. Find the solution to the equation.
$3(x+6)+2x=3x+5$

28. Solve the equation and check your solution.
$2\frac{6}{7}n-17=43$

Objective 3: Solve quadratic equations by factoring, extracting square roots, completing the square, and using the Quadratic Formula

29. Use the quadratic formula to solve the equation.
$2x^2+1=6x$

(A) $\frac{-3\pm\sqrt{7}}{2}$ (B) $\frac{3\pm\sqrt{7}}{2}$ (C) $\frac{-3\pm\sqrt{11}}{2}$ (D) $\frac{3\pm\sqrt{11}}{2}$

30. Solve the equation by factoring. (A) $-9,\ -4$ (B) $-4,\ 9$ (C) $4,\ 9$ (D) $-9,\ 4$
$x^2+5x-36=0$

31. Solve the equation by extracting square roots.
$(x+4)^2=1$

32. Use the quadratic formula to solve the equation.
$2x^2 - 8x + 3 = 0$

Objective 4: Solve polynomial equations of degree three or greater

Find all real solutions of the equation. Check your solutions in the original equation.

33. $35x^4 - 58x^2 + 24 = 0$

(A) $\pm 1,\ \pm 2$ (B) $\pm\sqrt{\frac{6}{7}},\ \pm\sqrt{\frac{4}{5}}$ (C) $\pm\sqrt{\frac{6}{5}},\ \pm\sqrt{\frac{4}{7}}$ (D) $\pm\sqrt{\frac{1}{3}},\ \pm\sqrt{\frac{5}{6}}$

34. $2x^4 - 5x^3 - 15x^2 + 10x + 8$

(A) $(2x+1)(x+4)(x+2)(x+1)$ (B) $(2x-1)(x-4)(x-2)(x-1)$
(C) $(2x+1)(x-4)(x+2)(x-1)$ (D) $(2x-1)(x+4)(x-2)(x+1)$

35. $x^3 - 5x^2 - x + 5 = 0$

36. $8x^3 + 343 = 0$

Objective 5: Solve equations involving radicals

Find all real solutions of the equation. Check your solutions in the original equation.

37. $\sqrt{x-8} - 7 = -3$ (A) –2 (B) –48 (C) 24 (D) no solution

38. $(x+3)^{4/3} = 81$ (A) –30 (B) –30 or 24 (C) 24 or 30 (D) 24

39. $3x(x+18)^{1/2} + 3(x+18)^{3/2} = 0$

40. $\sqrt{3x-5} = \sqrt{2x-3}$

Objective 6: Solve equations involving absolute values

Find all real solutions of the equation. Check your solutions in the original equation.

41. $|x^2+4| = -6x-5$ (A) 4 (B) –3, –2 (C) –8, –1 (D) –3

42. $5|x+3| - 25 = 10$ (A) 10, –4 (B) –3 (C) 4, –10 (D) no solution

43. $|x+3| = x^2-3$

44. $|3x+1| = 4$

Section P.3: The Cartesian Plane and Graphs of Equations

Objective 1: Plot points in the Cartesian plane

45. Identify the graph of the points $A(-4, -2)$, $B(3, 1)$, and $C(4, -1)$.

(A)

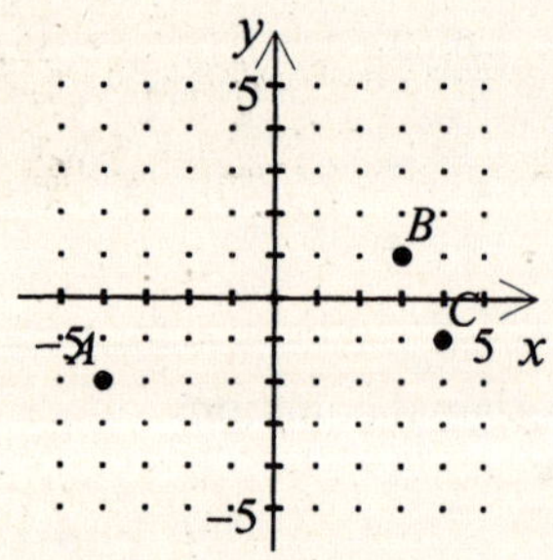

(B)

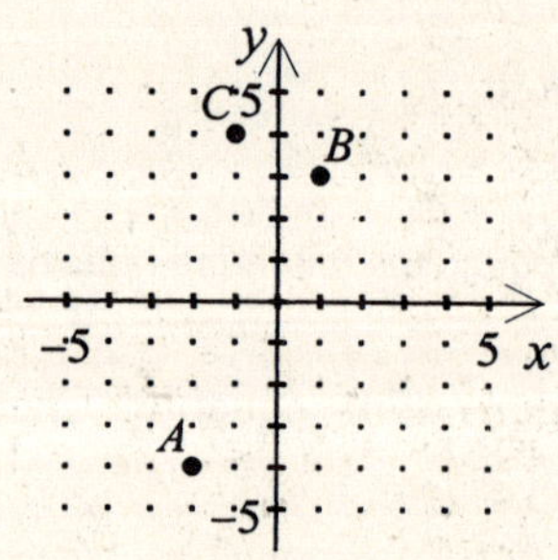

(C)

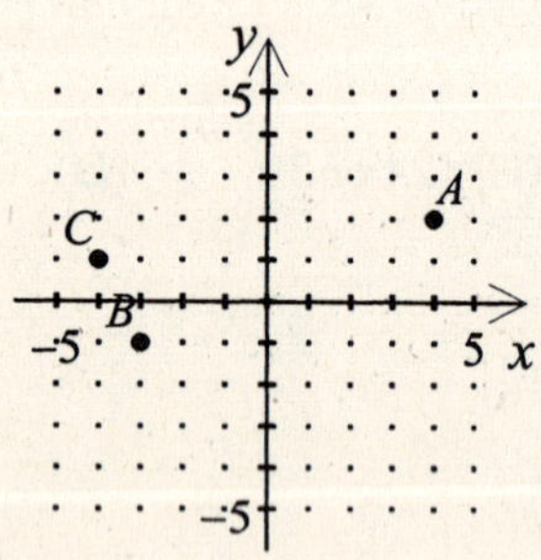

(D)

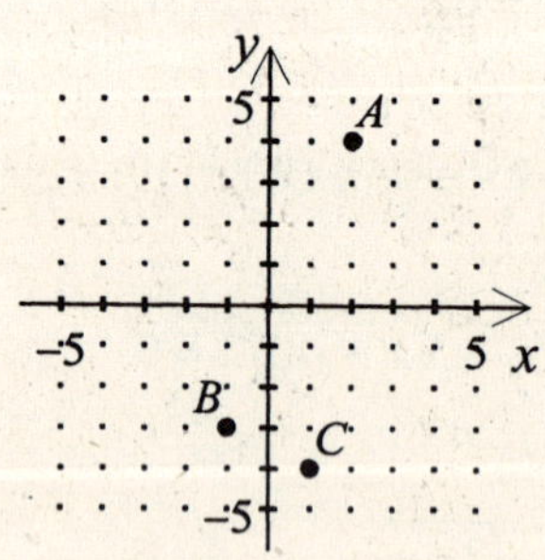

46. Determine the quadrant in which (x, y) is located so that the conditions are satisfied.

$x < 0$ and $y = -9$

(A) Quadrant I (B) Quadrant II (C) Quadrant III (D) Quadrant IV

47. Name the coordinates of point A and the quadrant in which A is located.

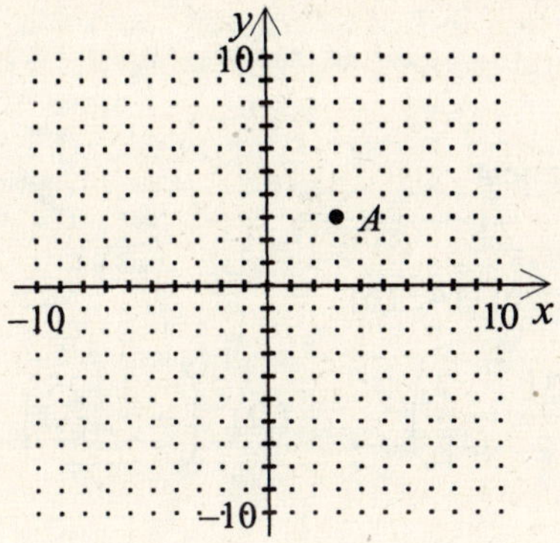

48. Name the point with coordinates $(5, 4)$.

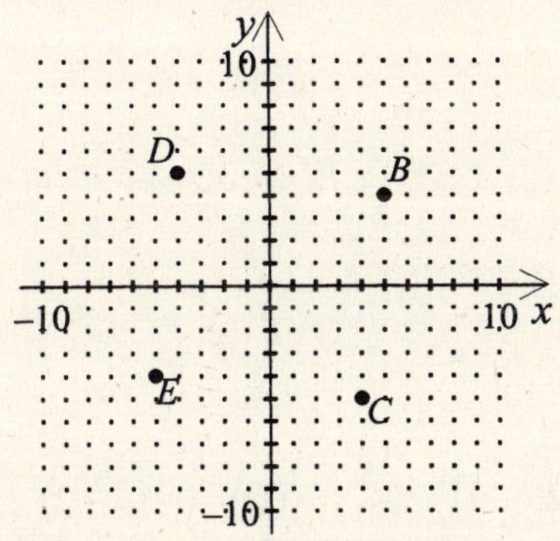

Objective 2: Use the Distance Formula to find the distance between two points

Find the distance between the points.

49. $(6, 3)$, $(-3, -4)$ (A) $\sqrt{130} \approx 11.40$ (B) 10 (C) 130 (D) $\sqrt{10} \approx 3.16$

50. $(4, -1)$, $(4, -4)$ (A) 3 (B) 89 (C) 9 (D) $\sqrt{89}$

Find the distance between the points.

51. $(-6, -2)$ and $(8, -2)$

52. Find the distance between the points $(-4.6, -4.7)$ and $(-3.4, -5)$. Round your answer to the nearest tenth.

Objective 3: Use the Midpoint Formula to find the midpoint of a line segment

53. Find the midpoint of the line segment connecting $(-17, 9)$ and $(14, 16)$.

(A) $(-3, 25)$ (B) $(3, -25)$ (C) $\left(-\frac{31}{2}, -\frac{7}{2}\right)$ (D) $\left(-\frac{3}{2}, \frac{25}{2}\right)$

54. Find the midpoint of $\overline{GH}$.

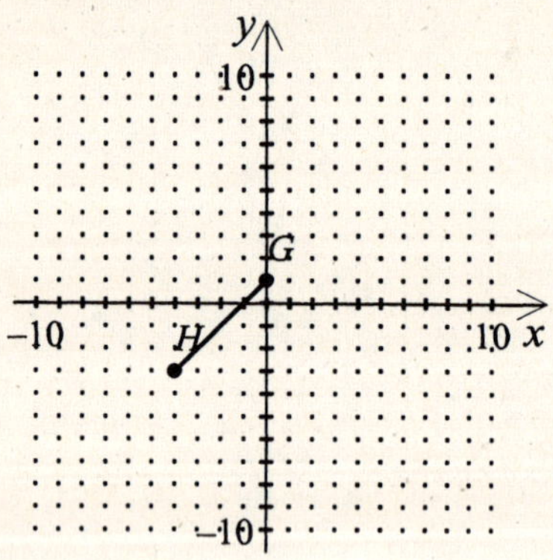

(A) $(2, 1)$ (B) $(1, 2)$ (C) $(-2, -1)$ (D) $(-1, -2)$

55. $M(2, 3)$ is the midpoint of $\overline{RS}$. If S has coordinates $(6, 6)$, find the coordinates of R.

56. Find the midpoint of the line segment connecting the two points. Then show that the midpoint is the same distance from each point.

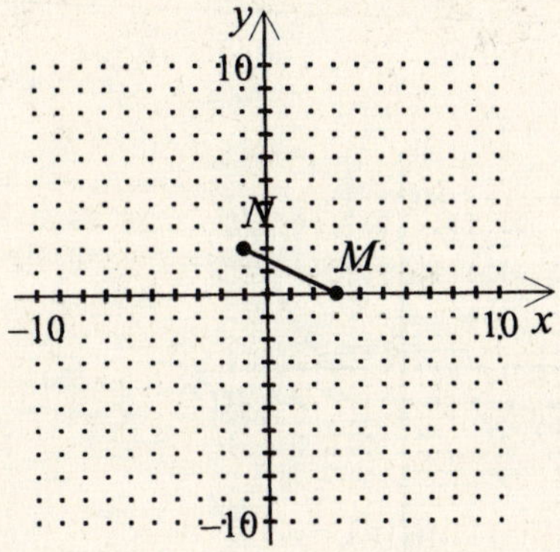

Objective 4: Sketch graphs of equations

Match the equation with its graph.

57. $y = x^3 + x^2 - 3x$

(A)

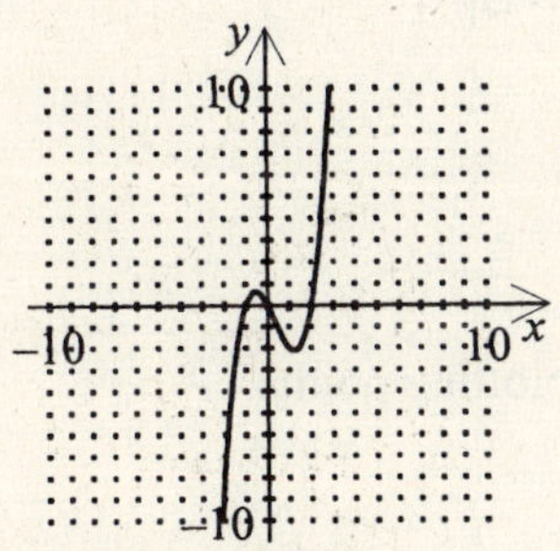

(B)

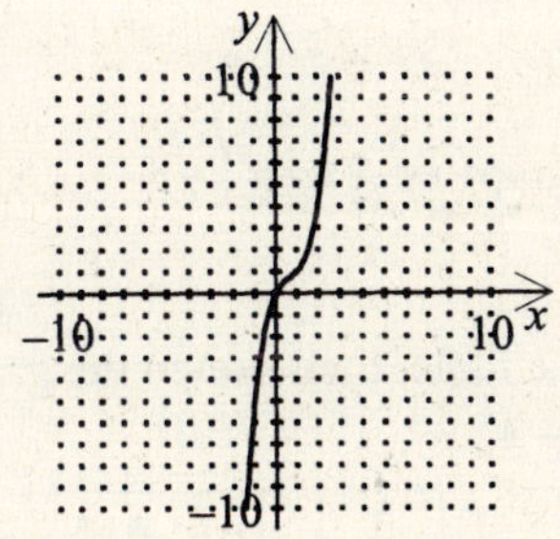

(C)

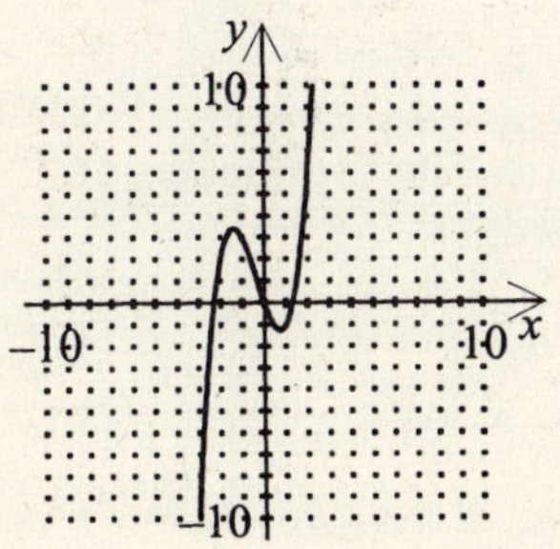

(D) none of these

Match the equation with its graph.

58. $y = -2\sqrt{x} + 3$

(A)

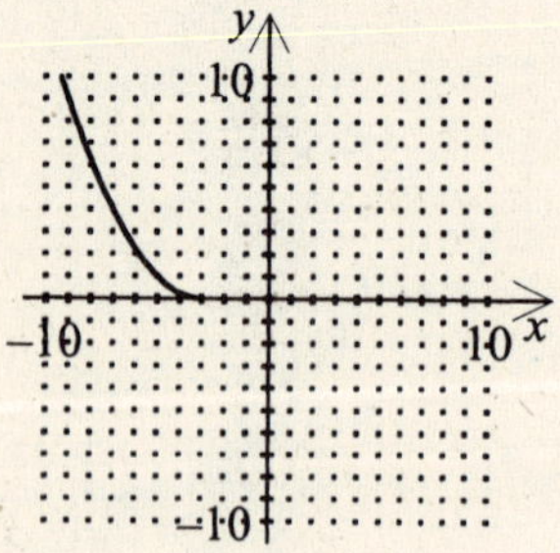

(B)

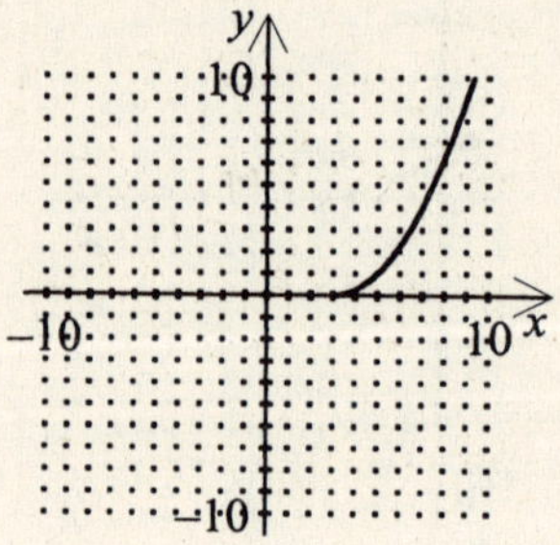

(C)

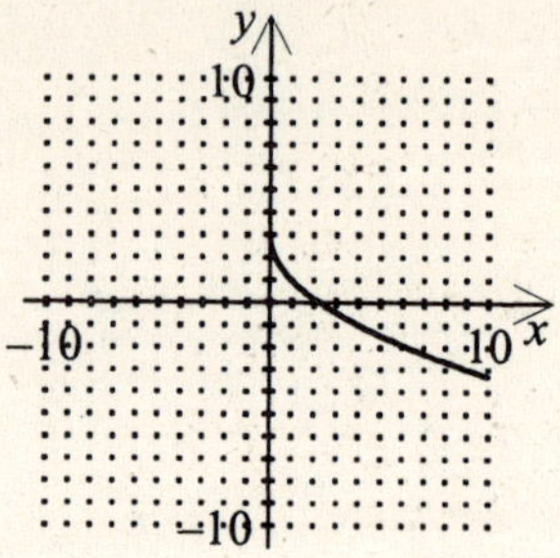

(D)

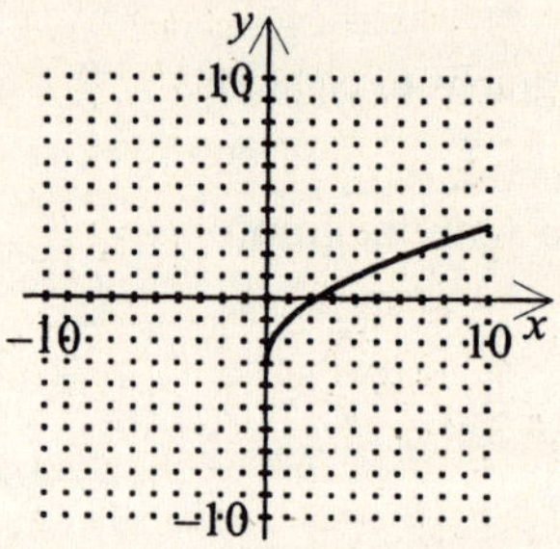

59. Sketch the graph of $y = |x| + 5$.

60. Complete the table, then sketch the graph of the equation by plotting points.
$-5x + y = -4$

x	-2	-1	0	1	2
y					

Objective 5: Find x- and y-intercepts of graphs

Identify the *x*- and *y*-intercepts of the graph of the equation.

61. $y = 2x^3\sqrt{9 - x^2}$

(A) *x*-intercepts: $(0,\ 0),\ \left(-\dfrac{9}{2},\ 0\right),\ \left(\dfrac{9}{2},\ 0\right)$
y-intercept: $(0,\ 0)$

(B) *x*-intercepts: $(3,\ 0),\ (-3,\ 0),\ (-9,\ 0)$
y-intercepts: none

(C) *x*-intercepts: $(0,\ 0),\ \left(-\dfrac{9}{2},\ 0\right),\ \left(\dfrac{9}{2},\ 0\right),\ (3,\ 0),\ (-3,\ 0)$
y-intercept: $(0,\ 0)$

(D) *x*-intercepts: $(0,\ 0),\ (3,\ 0),\ (-3,\ 0)$
y-intercept: $(0,\ 0)$

62. $y = \dfrac{2}{x-3} + 2$

(A) *x*-intercept: $\left(-\dfrac{1}{2},\ 0\right)$
y-intercept: $\left(0,\ -\dfrac{3}{4}\right)$

(B) *x*-intercept: $(-2,\ 0)$
y-intercept: $\left(0,\ -\dfrac{4}{3}\right)$

(C) *x*-intercept: $(2,\ 0)$
y-intercept: $\left(0,\ \dfrac{4}{3}\right)$

(D) *x*-intercept: $\left(\dfrac{1}{2},\ 0\right)$
y-intercept: $\left(0,\ \dfrac{3}{4}\right)$

63. $y = \sqrt{x+16} + 2$

64. $y = |\,x+2\,| + 4$

Objective 6: Use symmetry to sketch graphs of equations

65.

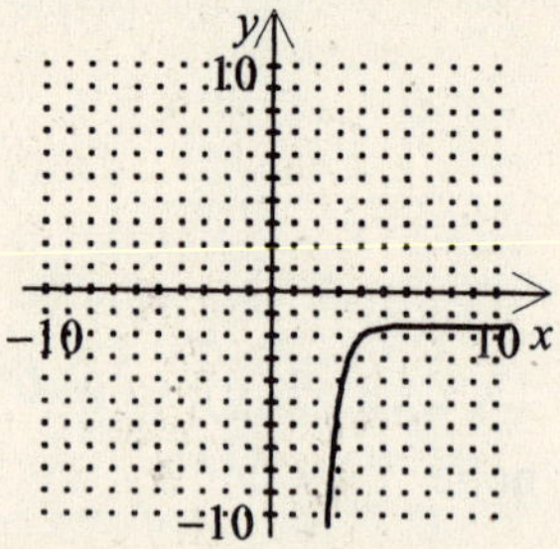

Assume that the graph above is a portion of a complete graph with y-axis symmetry. Which graph below is the complete graph?

(A)

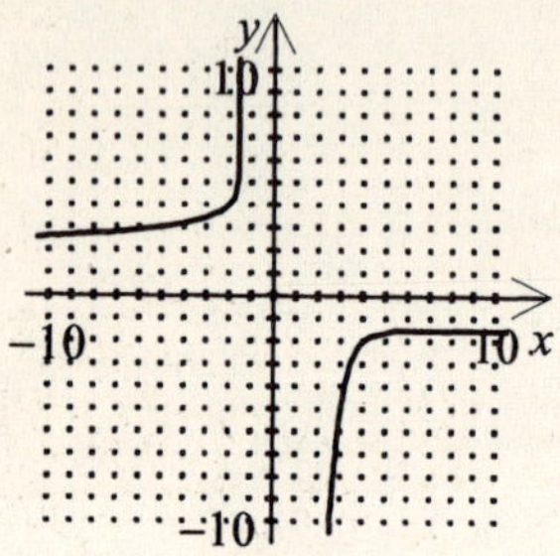

(B)

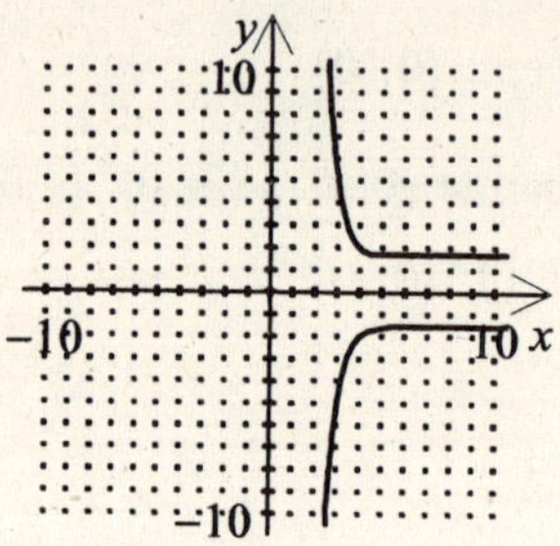

(C)

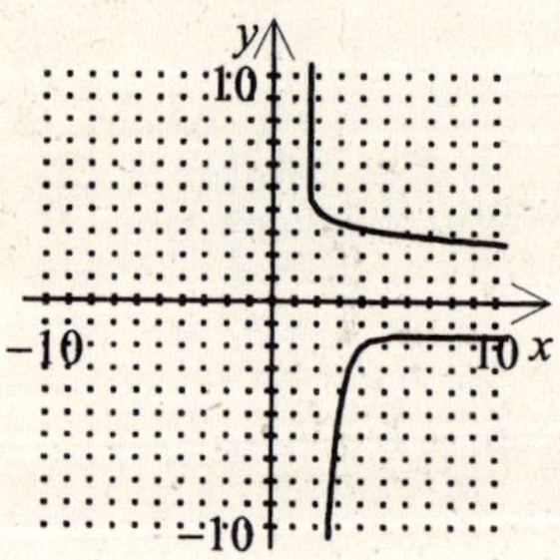

(D)

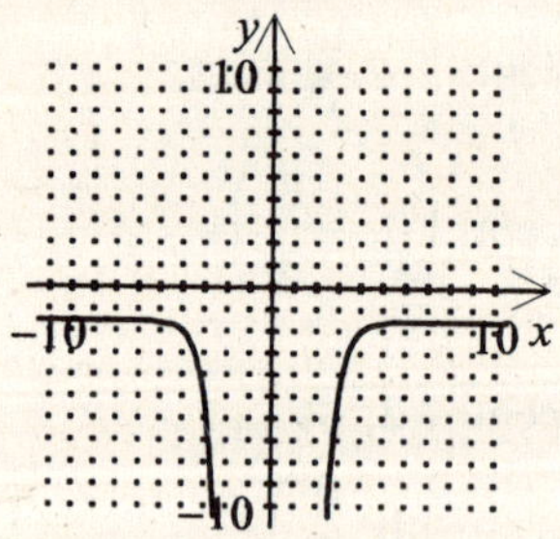

66. Identify the equation whose graph is symmetric with respect to the x-axis.

(A) $y^3 = 4x^4 + 4$ (B) $x = |y-3|$ (C) $4x^2 = y$ (D) $x = 2y^2 + 4$

67. The graph below is a portion of a complete graph. Sketch the complete graph assuming it is symmetric with respect to the x-axis.

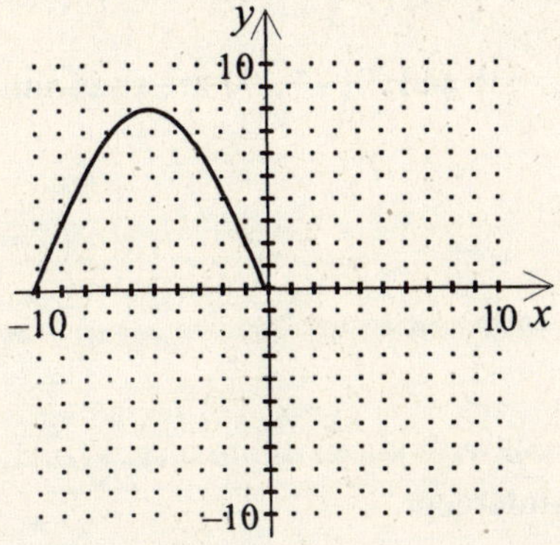

68. Determine whether the graph of $y = 4x^2 + 3$ is symmetric with respect to the x-axis, y-axis, and/or origin.

Objective 7: Find equations and sketch graphs of circles

69. Which equation does this graph represent?

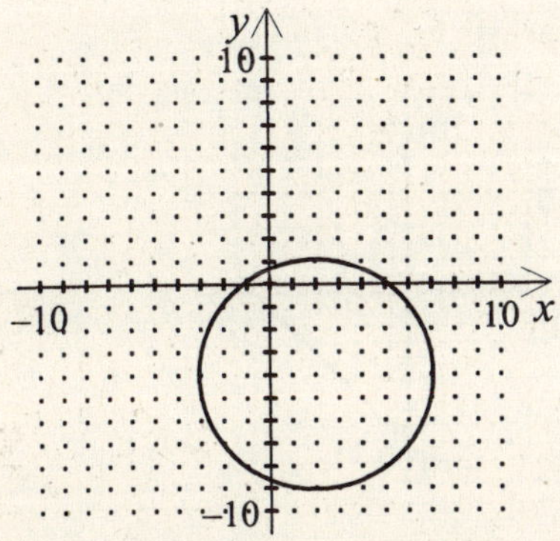

(A) $(x-2)^2 + (y+4)^2 = 25$ (B) $(x+2)^2 + (y-4)^2 = 5$

(C) $(x-4)^2 + (y-2)^2 = 5$ (D) $(x+5)^2 + (y-4)^2 = 25$

70. Find the center and radius of the circle with the given equation.
$(x-3)^2 + (y+4)^2 = 81$

(A) $(3,-4)$; 9 (B) $(-4,-3)$; 9 (C) $(-4,3)$; 81 (D) $(-3,4)$; 9

71. Sketch a graph of the equation $(x+1)^2+(y+2)^2=25$.

72. Find the equation in standard form of the circle where $C(5,-3)$ and $D(-7,7)$ are endpoints of a diameter.

Section P.4: Linear Equations in Two Variables

Objective 1: Use slope to graph linear equations in two variables

73. Identify the graph of the line that has the given slope and y-intercept.
$m=1$; y-intercept: $(0, 9)$

(A)

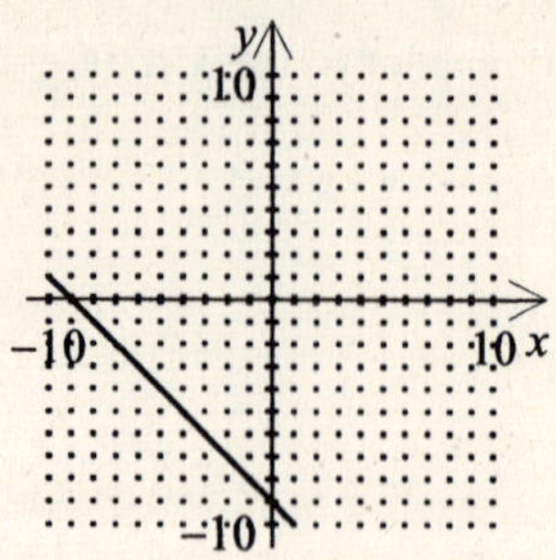

(B)

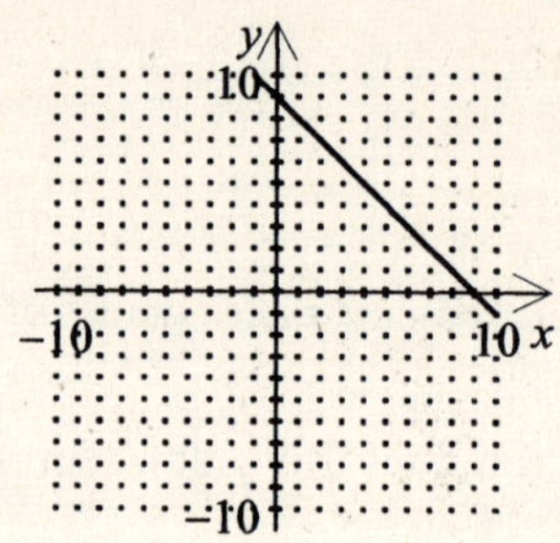

(C)

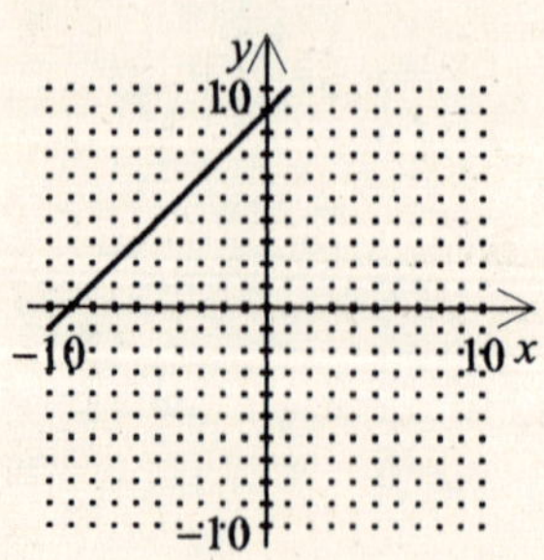

(D)

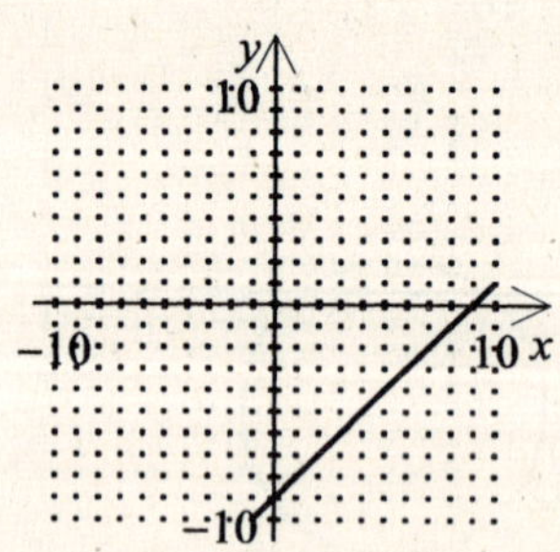

74. Identify the graph of the linear equation.
$4x = -16$

(A)
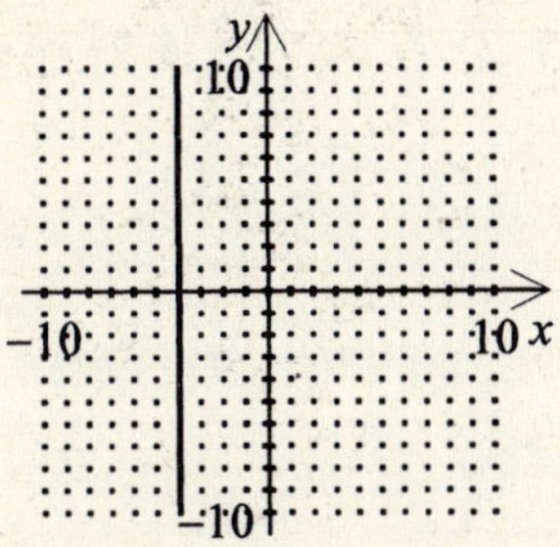

(B)
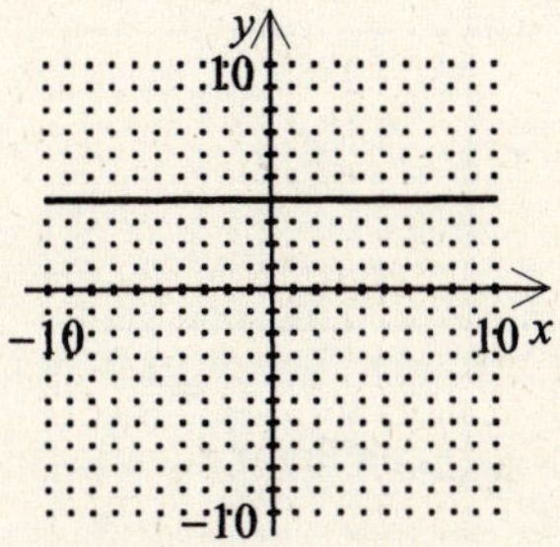

(C)
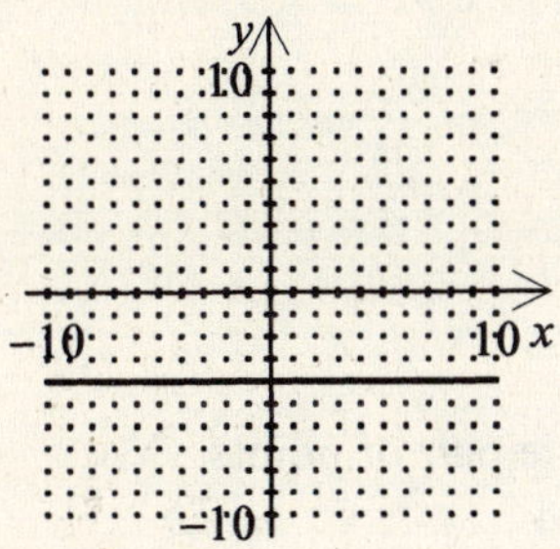

(D)
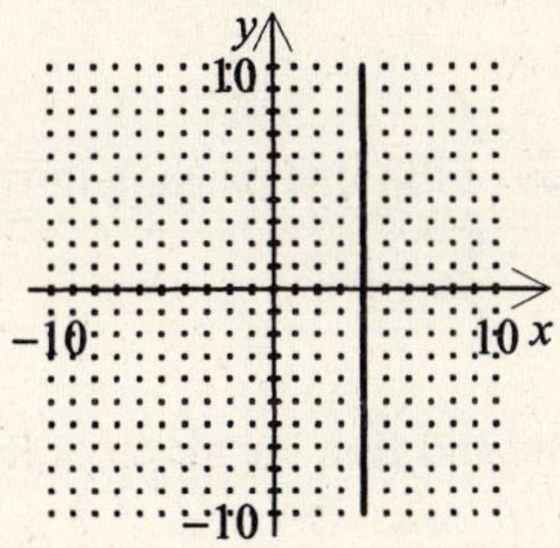

75. Use the point on the line and the slope of the line to find three additional points through which the line passes. (There are many correct answers.)
$(5, -1),\ m = 1$

76. Sketch the graph of the line through the given point with the indicated slope.
Point $(4, 4)$ with slope $\frac{3}{5}$

Objective 2: Find slopes of lines

77. Find the slope of the line passing through the pair of points.
$(-6, 4),\ (4, 4)$

(A) $\frac{4}{5}$ (B) 0 (C) –5 (D) Undefined

78. Estimate the slope of the line.

(A) -3 (B) $\frac{1}{3}$ (C) $-\frac{1}{3}$ (D) 3

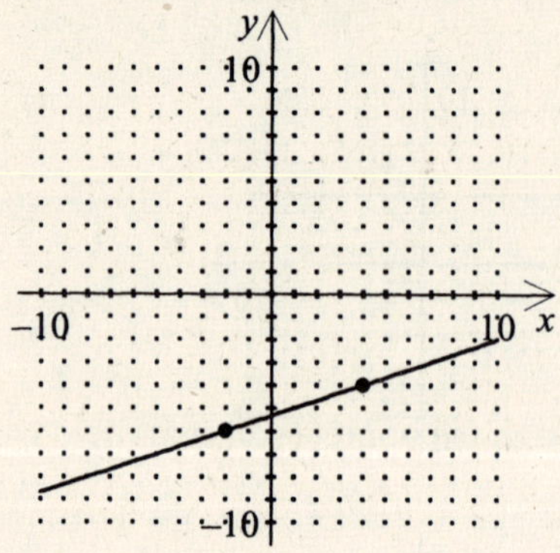

79. Find the slope of the line passing through the pair of points.
$(-8, -2),\ (-5, 3)$

80. Plot the points and find the slope of the line passing through the pair of points.
$\left(\frac{11}{3}, -\frac{10}{9}\right), \left(-\frac{4}{5}, -\frac{9}{4}\right)$

Objective 3: Write linear equations in two variables

81. Find the slope-intercept form of the equation of the line that passes through the given point and has the indicated slope.
$(-5, -3),\ m = -5$

(A) $y = 5x + 28$ (B) $y = -5x + 28$ (C) $y = -5x - 28$ (D) $y = 5x - 28$

Find the slope-intercept form of the equation of the line passing through the points.

82. $(7, 3),\ (-1, 9)$

(A) $y = -\frac{3}{4}x + \frac{33}{4}$ (B) $y = -\frac{3}{4}x + \frac{4}{33}$ (C) $y = -\frac{4}{3}x - \frac{33}{4}$ (D) $y = \frac{4}{3}x - \frac{4}{33}$

83. $(-5, 9),\ (-8, 9)$

84. Use the *intercept form* to find the equation of the line with the given intercepts. The intercept form of the equation of a line with intercepts $(a,\ 0)$ and $(0,\ b)$ is $\frac{x}{a}+\frac{y}{b}=1,\ a\neq 0,\ b\neq 0$.

x-intercept: $(-4,\ 0)$

y-intercept: $(0,\ 6)$

Objective 4: Use slope to identify parallel and perpendicular lines

85. Find the slope-intercept form of the equation of the line through the point $(6,\ 0)$, parallel to the line $-8x-5y=2$.

(A) $y=-\frac{8}{5}x-\frac{5}{48}$ (B) $y=-\frac{5}{8}x-\frac{5}{48}$ (C) $y=-\frac{8}{5}x+\frac{48}{5}$ (D) $y=\frac{8}{5}x+\frac{48}{5}$

86. Determine whether the graphs of the equations are parallel, perpendicular, or neither.

$3x-4y=8$

$x-4y=9$

(A) Parallel (B) Perpendicular (C) Neither

87. Find the slope-intercept form of the equation of the line that passes through the point $(2,\ 1)$ and is perpendicular to the line $4x-7y=-9$.

88. Determine whether the lines L_1 and L_2 passing through the pair of points are parallel, perpendicular, or neither.

L_1: $(3,\ 10),\ (5,\ 7)$

L_2: $(-10,\ 8),\ (-8,\ 5)$

Objective 5: Use linear equations in two variables to model and solve real-life problems

89. In 1980, the average price of a home in Lake County was \$97,000. By 1985, the average price of a home was \$112,000. Which of the following is a linear model for the price P of a home in Lake County, in terms of the year t? Let $t=0$ correspond to 1980.

(A) $P=15{,}000t+97{,}000$ (B) $P=3000t+97{,}000$

(C) $P=112{,}000-3000t$ (D) $P=112{,}000-15{,}000t$

90. The graph for a stable that charges a $20 flat fee plus $10 per hour for horseback riding is shown below. How will the graph change if the stable changes its charges to a flat fee of $40 plus $20 per hour?

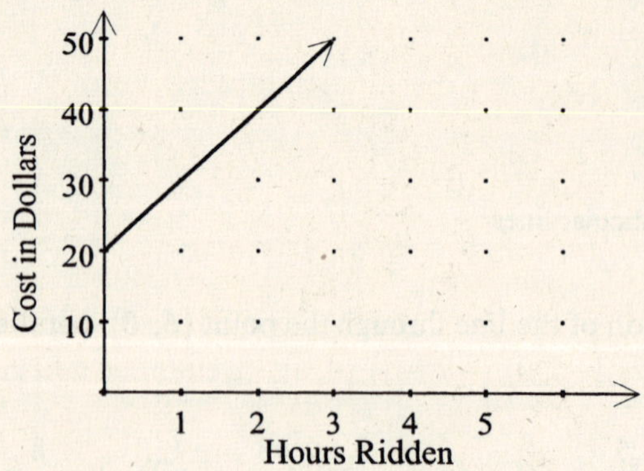

(A) The slope will be 40 and the y-intercept will be 20.

(B) The slope will be 20 and the y-intercept will be 40.

(C) The slope will be 10 and the y-intercept will be 40.

(D) The slope will be 20 and the y-intercept will be 20.

91. The graph below shows the weights of several kittens at various ages.

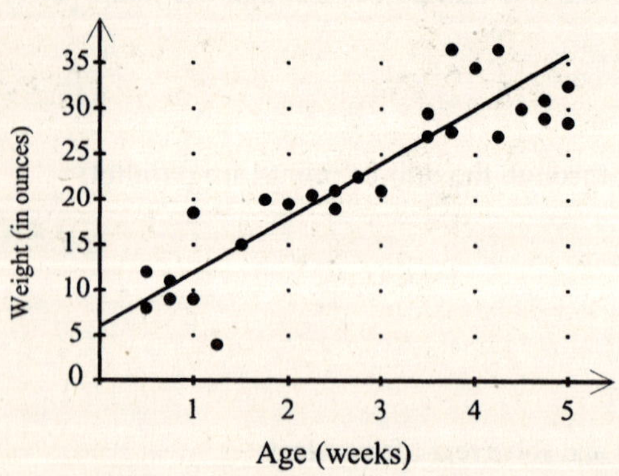

(a) What is an equation of the line of best fit?
(b) What is the meaning of the slope and y-intercept of the graph of the line of best fit?

92. Write an equation for the following situation in slope-intercept form.

The gas tank in a truck holds 13 gallons. The truck uses $\frac{2}{5}$ gallon per mile.

Section P.5: Functions

Objective 1: Determine whether relations between two variables are functions

93. Determine which set of ordered pairs (x, y) represents y as a function of x.

(A) $\{(-6, -4), (-4, 0), (-6, 4)\}$ (B) $\{(-6, -4), (0, 4), (0, -6), (4, 0)\}$

(C) $\{-6, -4, 0, 4\}$ (D) $\{(-6, -4), (-4, -6), (4, 4)\}$

94. Identify the model of the relation $\{(-9,-2),(-5,0),(-9,-1),(4,6)\}$. Determine whether the relation is a function.

(A)

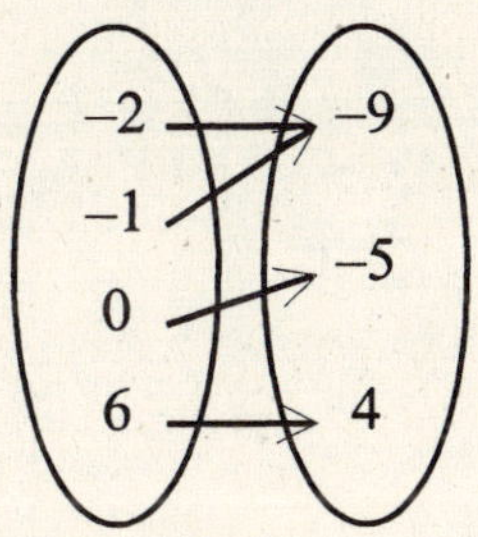

Not a function

(B)

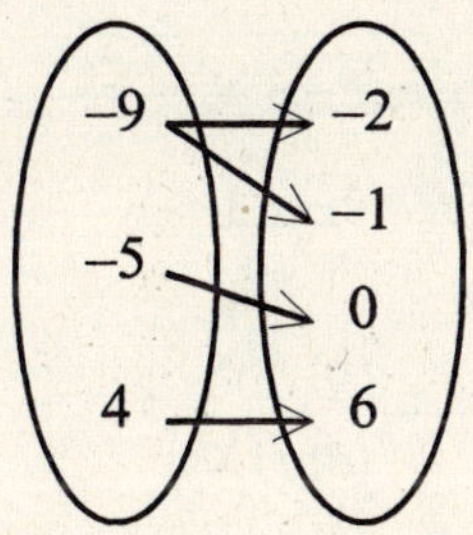

Function

(C)

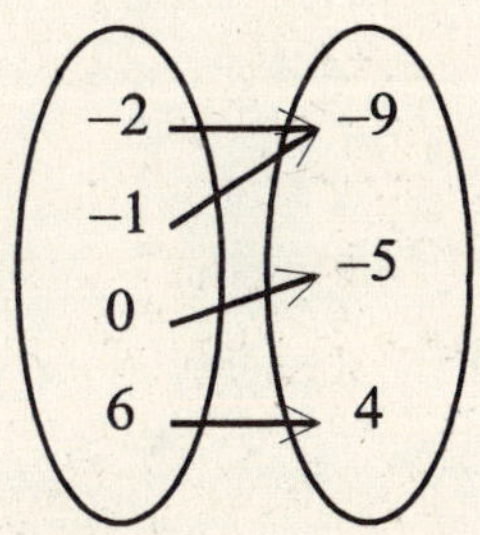

Function

(D)

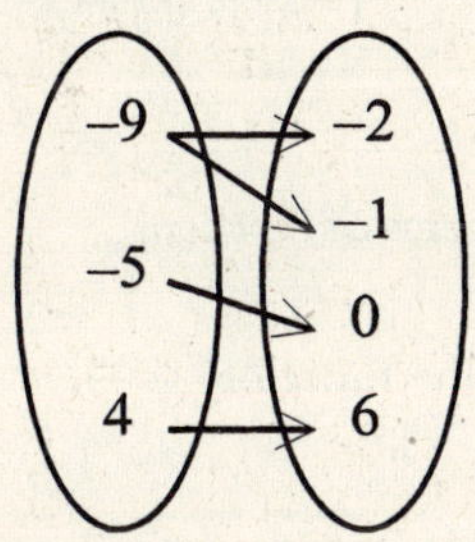

Not a function

95. Does the table describe y as a function of x? Explain your reasoning.

Input x	1	2	3	4
Output y	1	4	9	16

96. Determine whether the equation represents y as a function of x.

$x^2 + y^2 = 4$

Objective 2: Use function notation and evaluate functions

Evaluate the function at the specified value(s) of the independent variable and simplify.

97. $f(x)=8x^2-\sqrt{4x}\ ;\ f(6)$ (A) 278.202 (B) 276 (C) 264 (D) 283.101

98. $g(x)=\dfrac{x^2-7}{3x};\ g(n-2)$

(A) $\dfrac{n^2-4n+-3}{3n-2}$ (B) $\dfrac{n^2-9}{3n-6}$ (C) $\dfrac{n^2-7}{3n}-2$ (D) $\dfrac{n^2-4n+-3}{3n-6}$

99. $f(x)=\begin{cases}\dfrac{1}{6}x & \text{if } x>3\\ 8-x^2 & \text{if } x\le 3\end{cases}$

(a) $f(3)$ (b) $f(0)$
(c) $f(1)$ (d) $f(2.8)$

100. $f(x)=\dfrac{7}{9}x+9;\ f(-27),\ f(54)$

Objective 3: Find domains of functions

Find the domain of the function.

101. $h(x)=\dfrac{9x}{x\left(x^2-64\right)}$

(A) All real numbers $x\neq\pm 64, 0$ (B) All real numbers $x\neq 8$

(C) All real numbers $x\neq\pm 8, 0$ (D) All real numbers $x\neq\pm 8$

102. $f(x)=\sqrt{-x+7}$ (A) $x\le 7$ (B) $x\ge 0$ (C) $x\ge 7$ (D) $x\le 0$

103. $f(x)=\dfrac{\sqrt{x+10}}{x^2-6x+8}$

Find the domain of the function.

104. $f(x)=\sqrt[6]{x^7}$

Objective 4: Use functions to model and solve real-life problems

105. A slug travels at a rate of 0.6 meter per minute. Let t be the number of minutes spent traveling.
(a) Find the distance d traveled by the slug as a function of t.
(b) Find the distance traveled by the slug in 3 minutes.
(c) Identify the graph illustrating the distance traveled by the slug versus time.

(A) (a) $d(t)=t+0.6$
(b) 2.1 m
(c)
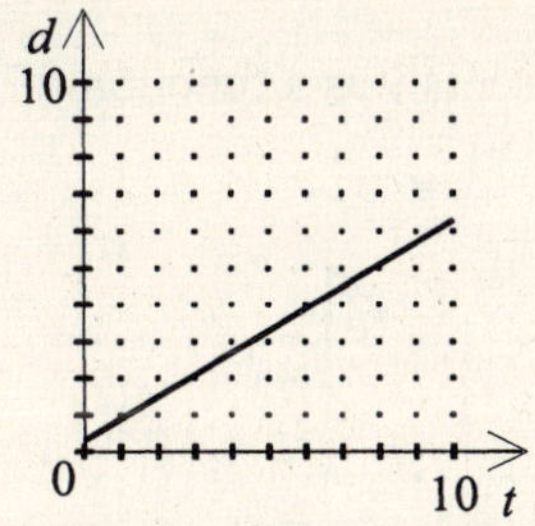

(B) (a) $d(t)=0.6d$
(b) 1.6 m
(c)
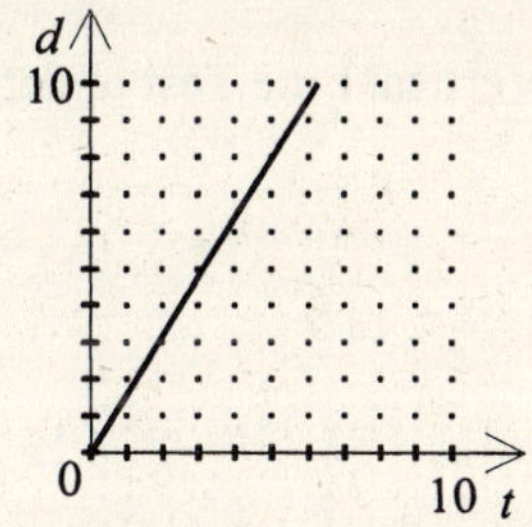

(C) (a) $d(t)=0.6t$
(b) 1.8 m
(c)
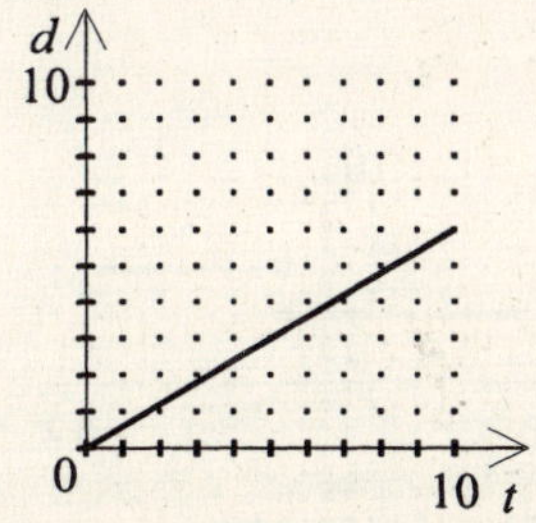

(D) (a) $d(t)=0.6t+1$
(b) 1.7 m
(c)
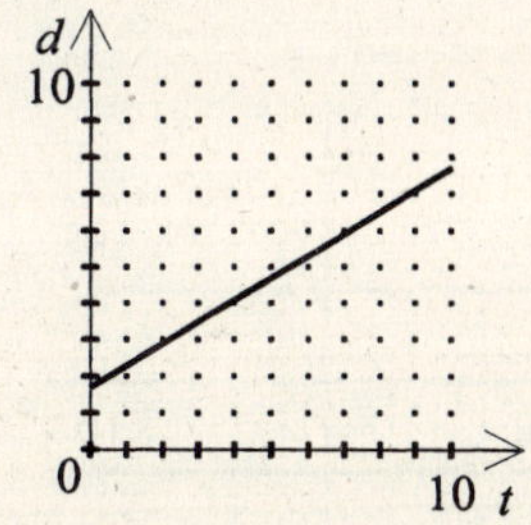

106. A publishing company estimates that the average cost (in dollars) for one copy of a new scenic calendar it plans to produce can be approximated by the function
$C(x)=\dfrac{4.25x+600}{x}$
where x is the number of calendars printed. Find the average cost per calendar when the company prints 10,000 calendars.

(A) \$64.25 (B) \$10.25 (C) \$4.31 (D) \$0.06

107. Since 1993, Rebecca Acosta has owned a bookstore called The Fly Leaf. The number of books B, in thousands, that The Fly Leaf has sold each year can be modeled by the function

$$B(t) = t^2 + 29t + 400$$

where t is the number of years after 1993. Using this model, estimate the number of books sold in 1995.

108. A pepper plant in Jeremy's garden was 14 centimeters tall when it was first planted. Since then, it has grown approximately 0.8 centimeter per day. Write a function expressing the pepper plant's height H in terms of the number of days d since it was planted.

Section P.6: Analyzing Graphs of Functions

Objective 1: Use the Vertical Line Test for functions

109. Use the Vertical Line Test to determine which graph represents y as a function of x.

(A)

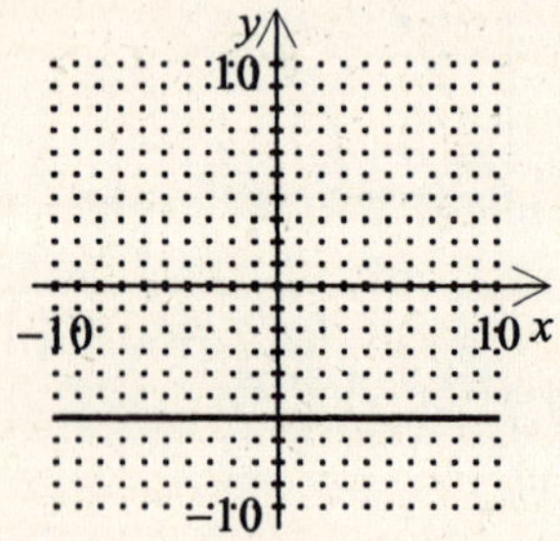

(B)

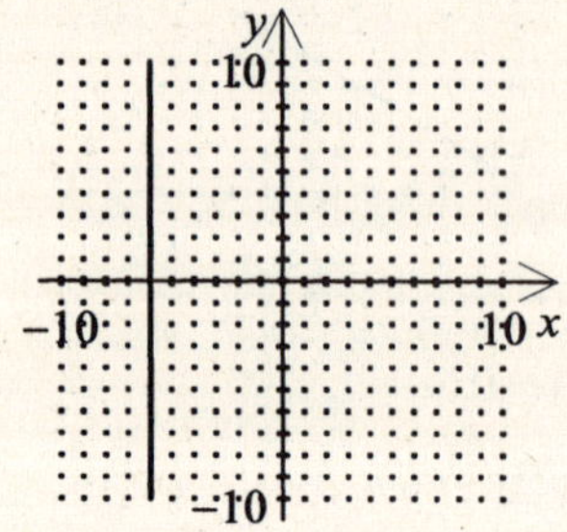

(C)

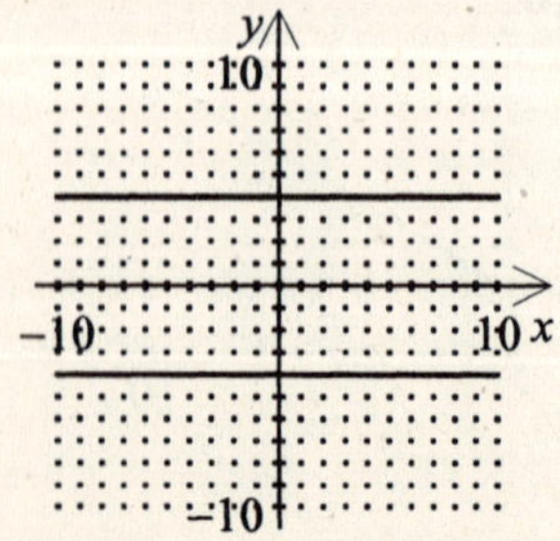

(D)

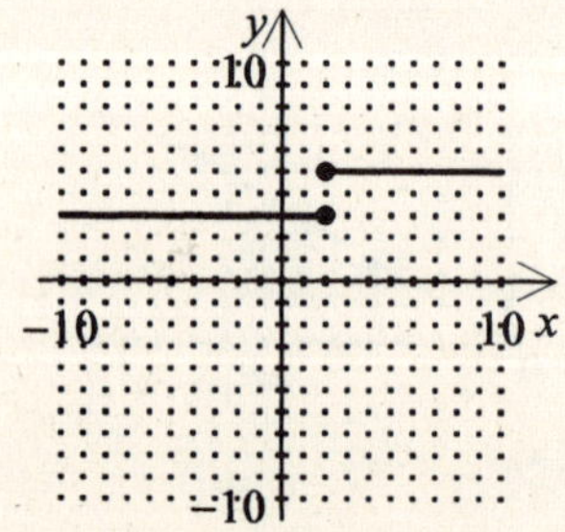

110. Use the Vertical Line Test to determine which graph represents y as a function of x.

(A)

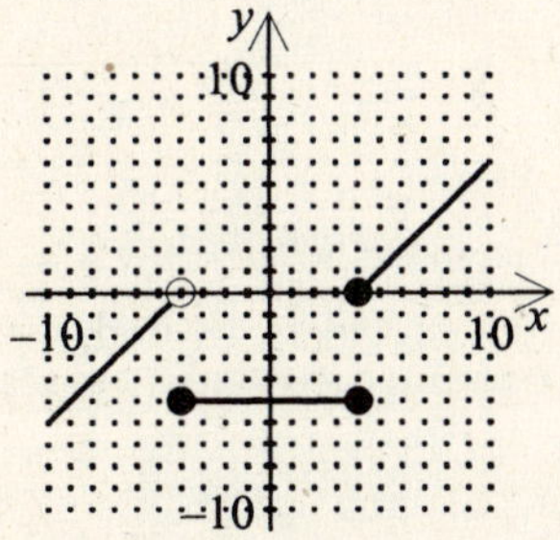

(B)

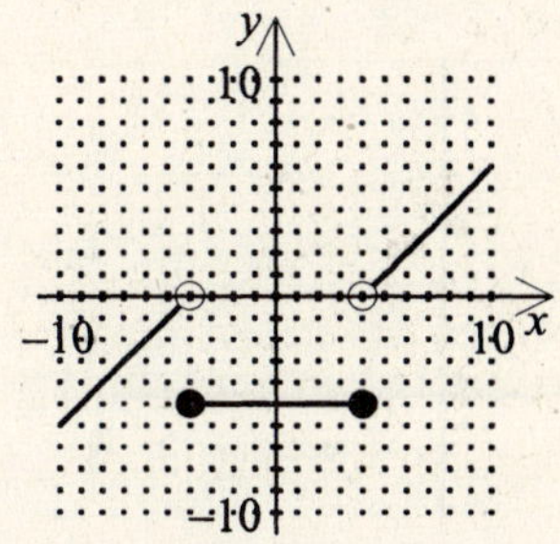

(C)

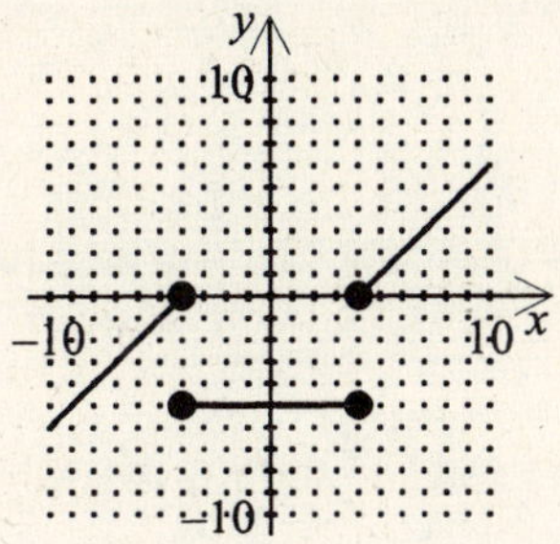

(D)

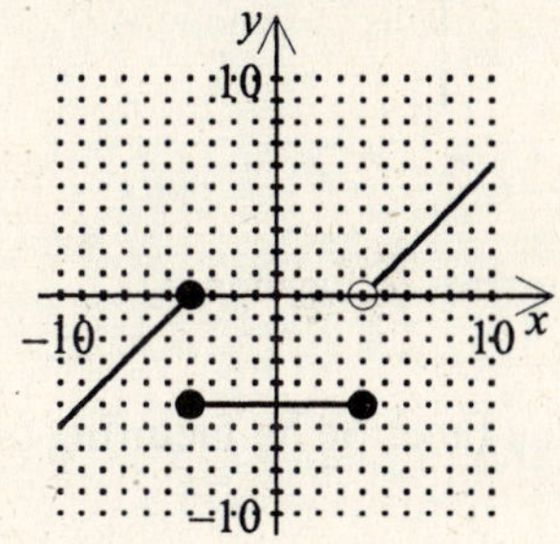

Use the Vertical Line Test to determine if the graph represents y as a function of x.

111.

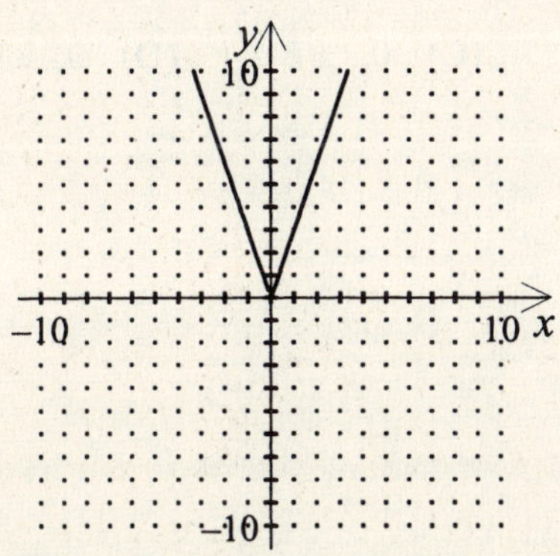

Use the Vertical Line Test to determine if the graph represents y as a function of x.

112.

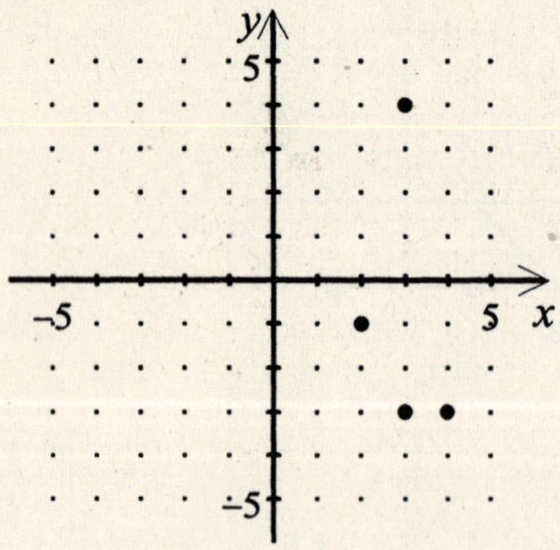

Objective 2: Find the zeros of functions

Find the zeros of the function by factoring or algebraically.

113. $f(x) = 2x^2 - 2x - 3$

(A) 0.82, –1.82 (B) 0.29, –1.29 (C) 1.29, –0.29 (D) 1.82, –0.82

114. $f(x) = \dfrac{x^3 - 36x}{(x+3)^2}$ (A) 0, ±3 (B) 36, ±6 (C) 0, ±6 (D) 0, –3, ±6

115. $f(x) = x + 2\sqrt{x} - 8$

116. $f(x) = -6x^4 + 30x^3 + 36x^2$

Objective 3: Determine intervals on which functions are increasing or decreasing

117. Use a graphing utility to graph the function and visually determine the intervals on which the function is increasing, decreasing, or constant.

$$f(x)=\begin{cases} -x^2, & x\le 2 \\ \sqrt{x-1}-5, & x>2 \end{cases}$$

(A)

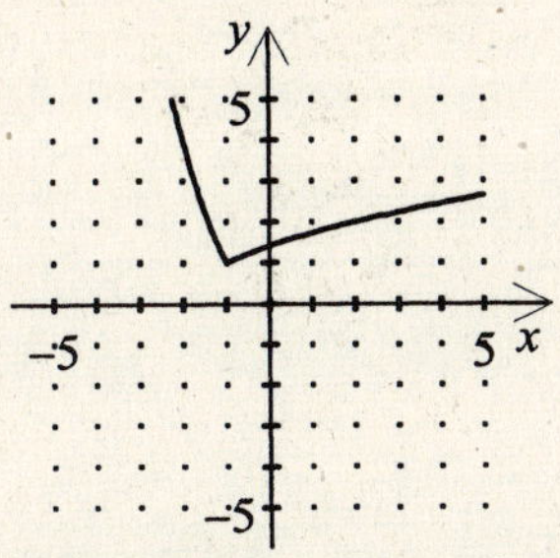

Increasing on $(-\infty,\ 0)$ and $(2,\ \infty)$

Decreasing on $(0,\ 2)$

(B)

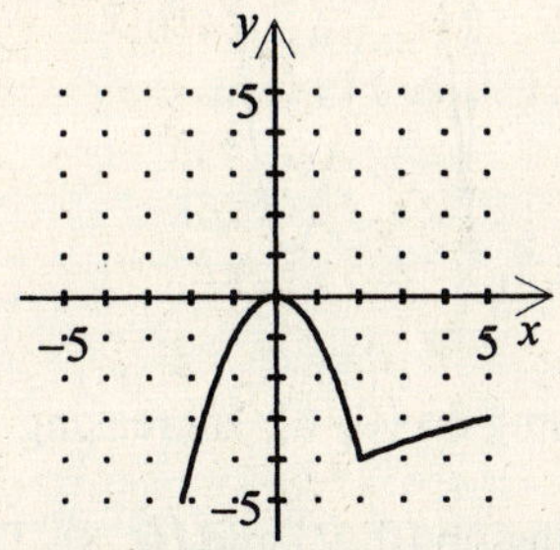

Increasing on $(-\infty,\ 0)$ and $(2,\ \infty)$

Decreasing on $(0,\ 2)$

(C)

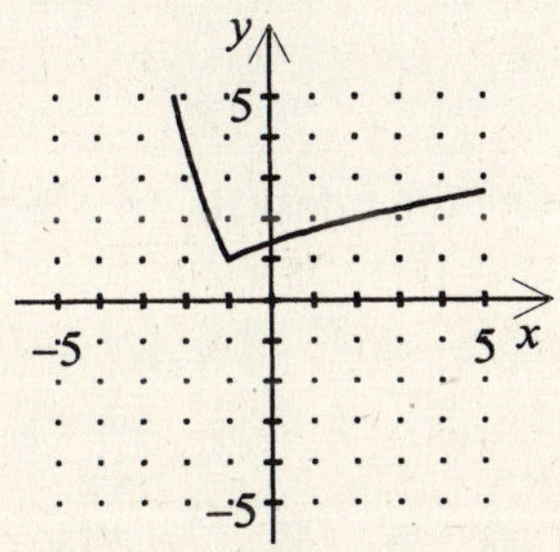

Decreasing on $(-\infty,\ 0)$ and $(2,\ \infty)$

Increasing on $(0,\ 2)$

(D)

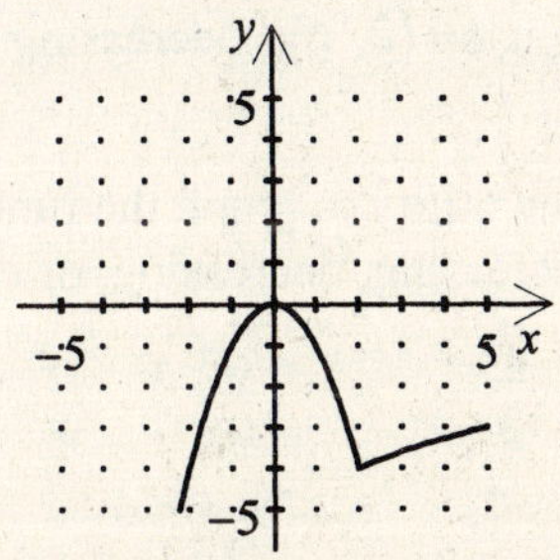

Decreasing on $(-\infty,\ 0)$ and $(2,\ \infty)$

Increasing on $(0,\ 2)$

118. Determine the intervals on which the function is increasing, decreasing, or constant.

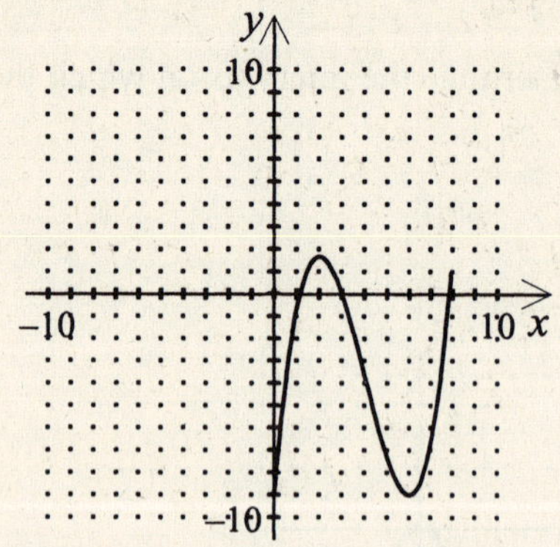

(A) Decreasing on $(0,\ 6)$; Increasing on $(6,\ 8)$

(B) Increasing on $(0,\ 2)$ and $(6,\ 8)$; Decreasing on $(2,\ 6)$

(C) Decreasing on $(0,\ 2)$ and $(6,\ 8)$; Increasing on $(2,\ 6)$

(D) Increasing on $(0,\ 6)$; Decreasing on $(6,\ 8)$

119. Use a graphing utility to graph the function and visually determine the intervals on which the function is increasing, decreasing, or constant.

$$f(x)=\begin{cases}\dfrac{7}{3}x, & x\le 3\\ 7, & 3<x\le 6\\ -\dfrac{1}{3}x+9, & x>6\end{cases}$$

120. (a) Use a graphing utility to graph the function and visually determine the intervals on which the function is increasing, decreasing, or constant.
(b) Make a table of values to verify whether the function is increasing, decreasing, or constant.
$f(x)=-\sqrt[3]{x-5}$

Objective 4: Identify even and odd functions

121. Identify the function that is *neither* odd nor even.

(A) $f(x)=2x^3+7x$

(B) $f(x)=2x^4-|x^4|-4$

(C) $f(x)=x^6+7x+4+|x-2|$

(D) $f(x)=x^6+7x^4+2x^2$

122. Identify the function that is neither odd nor even.

(A) $p(x) = -5x^3 + 5x - 4$

(B) $h(x) = -5x^4 + 5x^2 + 3$

(C) $g(x) = 3x^4 + 5x^2$

(D) $f(x) = \dfrac{-5x^3}{-5x^2 + 5}$

123. Determine whether the function is even, odd, or neither.
$f(x) = -6x^3 - 6x - 6$

124. Use the graph to determine if the function is even, odd, or neither.

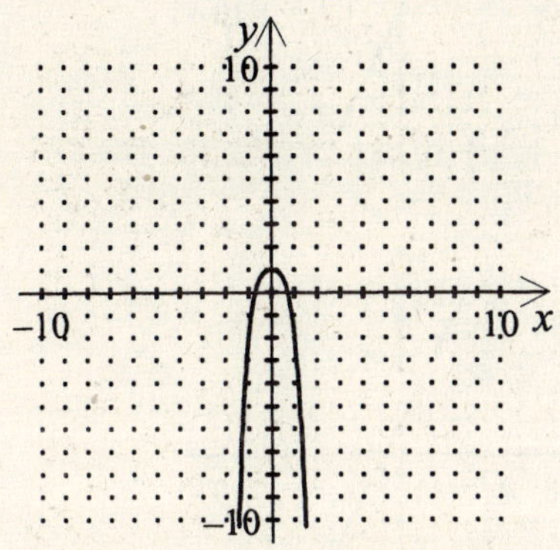

Section P.7: A Library of Functions

Objective 1: Identify and graph linear and squaring functions

125. Identify the graph of the linear function.
$f(x) = -2x + 2$

(A)

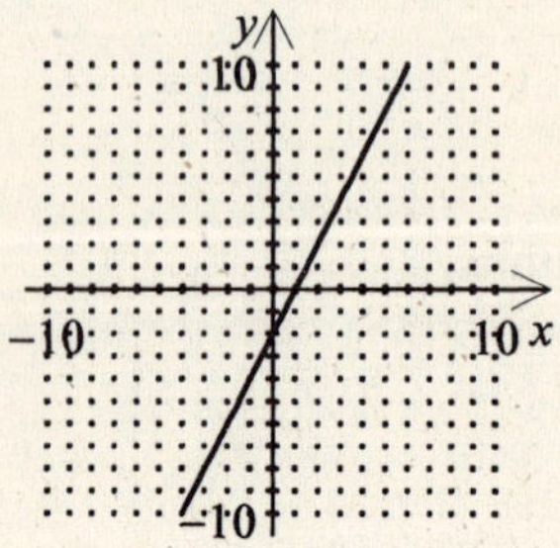

(B)

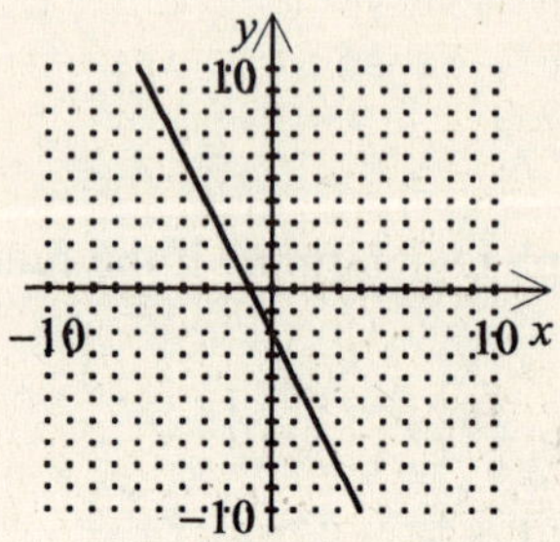

(C)

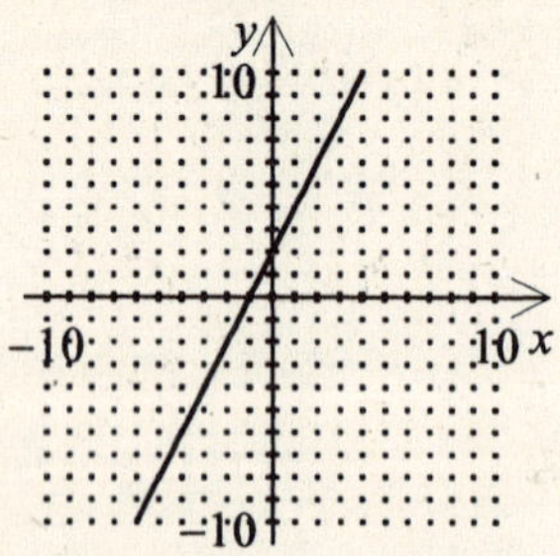

(D)

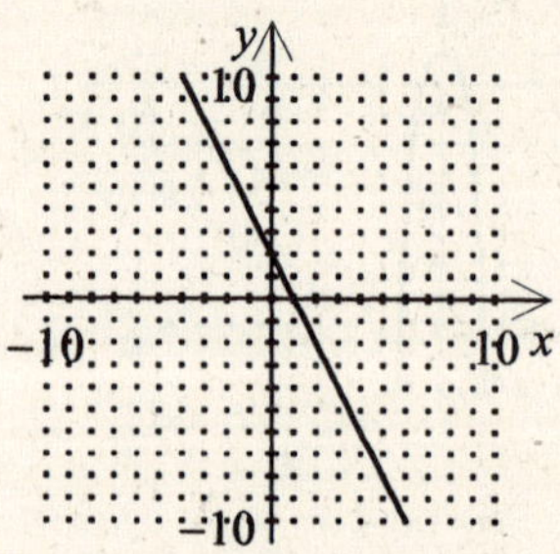

126. Identify the graph of the function and find the y-intercept.
$f(x) = 0.4x + 3$

(A)

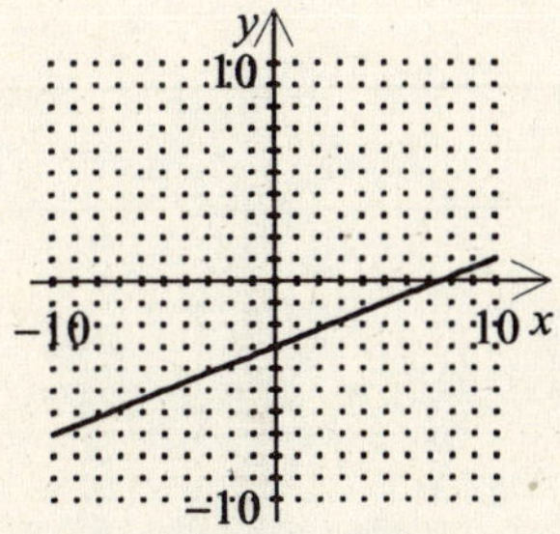

y-intercept: $(0, -3)$

(B)

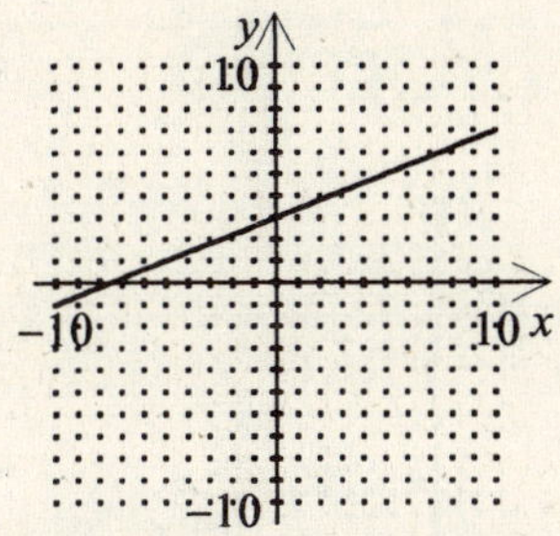

y-intercept: $(0, 3)$

(C)

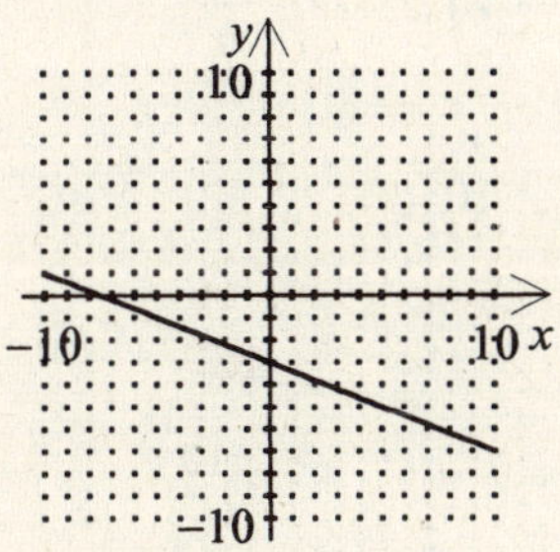

y-intercept: $(0, -3)$

(D)

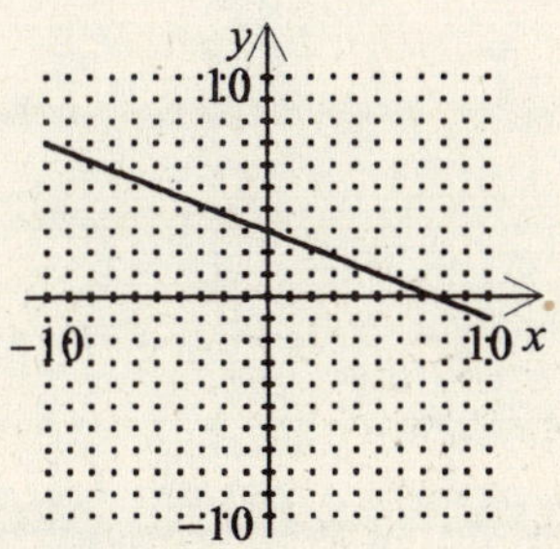

y-intercept: $(0, 3)$

(126.)

127. Sketch the graph of the linear function.

$$f(x) = -\frac{3}{2}x - 1$$

128. Write the linear function that has the indicated function values.

$f(4) = 2$, $f(9) = 7$

Objective 2: Identify and graph cubic, square root, and reciprocal functions

129. Which could be the graph of $y = \frac{3}{5}\sqrt{x}$?

(A)

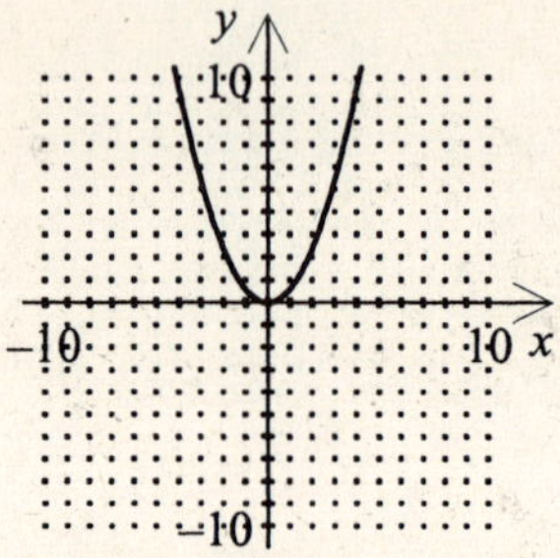

(B)

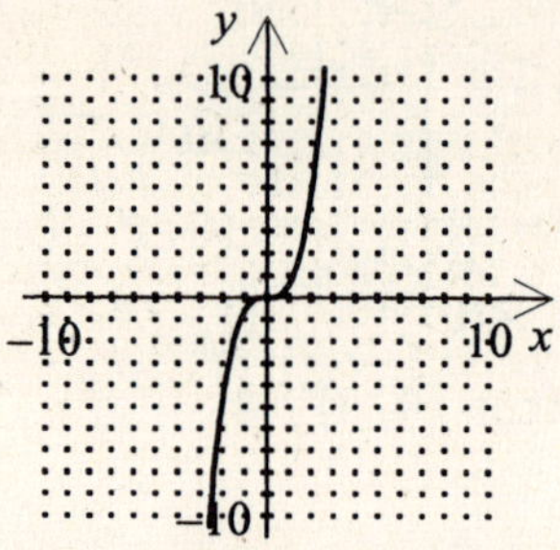

(C)

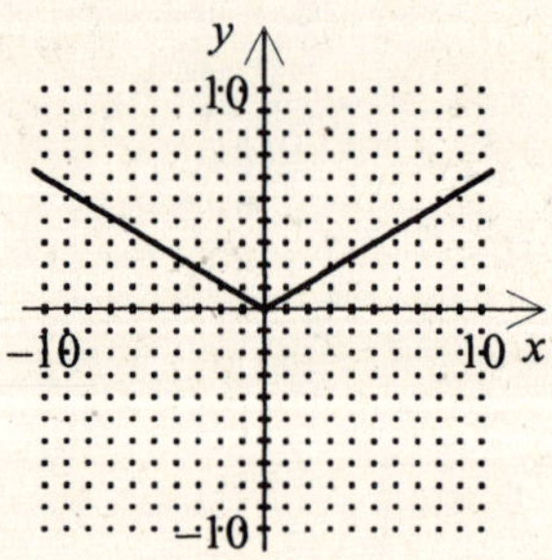

(D)

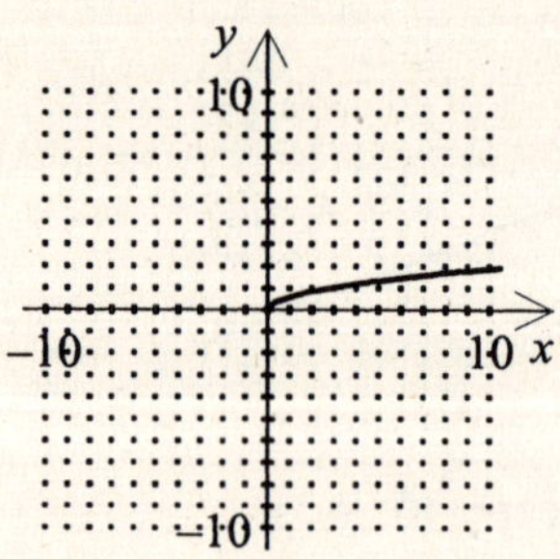

130. Which could be the graph of $y = \frac{1}{x}$?

(A)

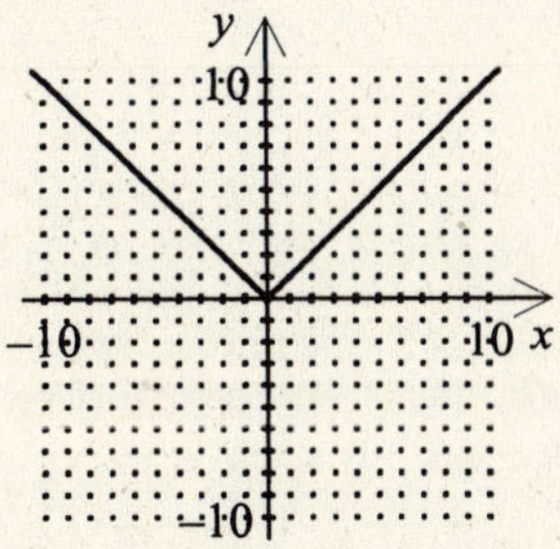

(B)

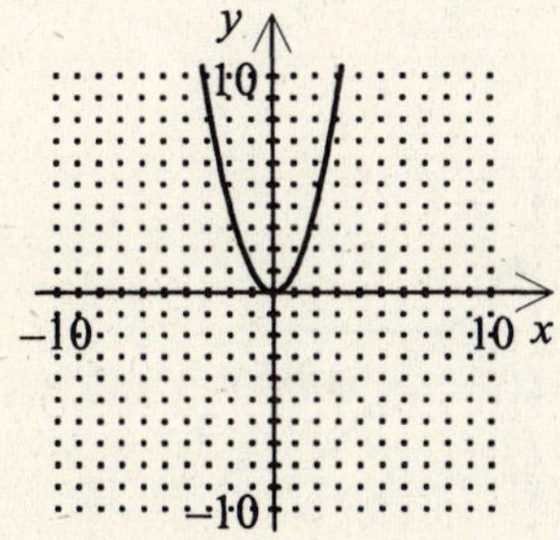

(C)

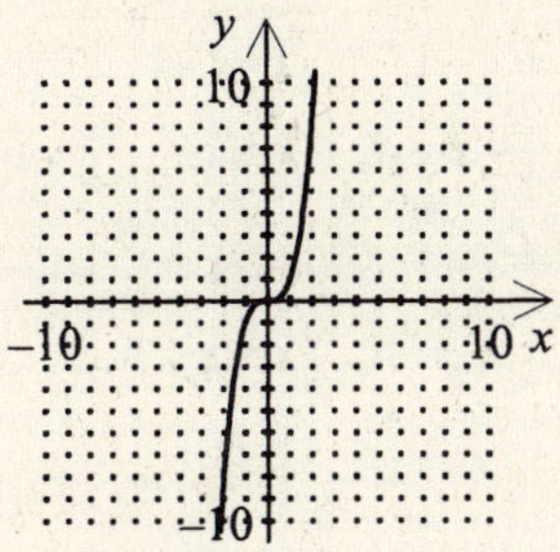

(D)

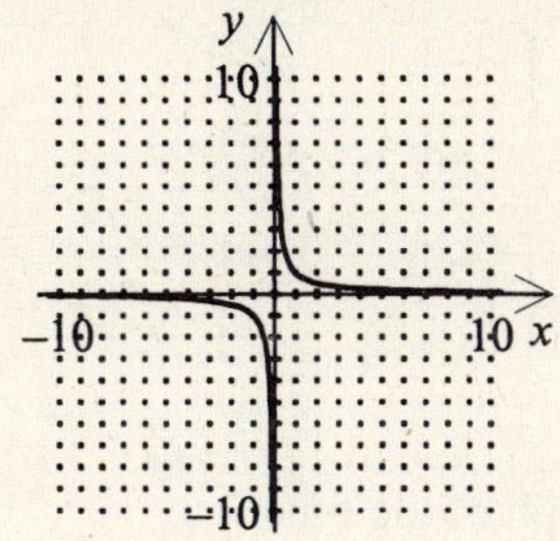

131. List an intercept of the graph of the square root function.

132. What is the name given to the function $y = \frac{1}{x}$?

Objective 3: Identify and graph step and other piecewise-defined functions

133. Find the function that represents the graph.

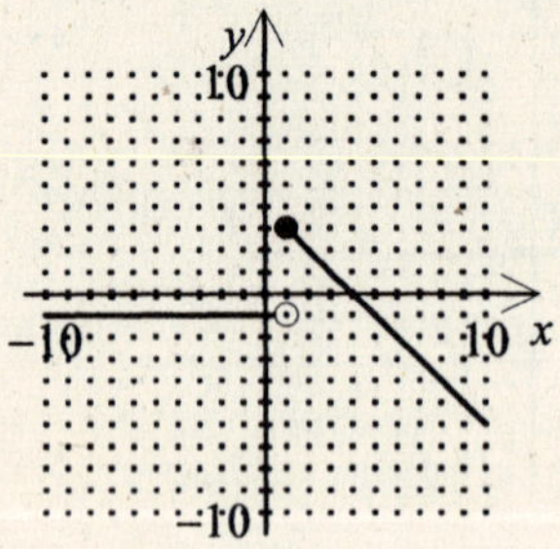

(A) $f(x)=\begin{cases}-1, & x<1\\ -x+4, & x\geq 1\end{cases}$

(B) $f(x)=\begin{cases}-1, & x<1\\ -x+4, & x\leq 1\end{cases}$

(C) $f(x)=\begin{cases}-1, & x>1\\ -x+4, & x\geq 1\end{cases}$

(D) $f(x)=\begin{cases}-1, & x>1\\ -x+4, & x\leq 1\end{cases}$

134. Find the graph of the function.

$$f(x)=\begin{cases}\sqrt{x}+4, & x\geq 0\\ -x+1, & x<0\end{cases}$$

(A)

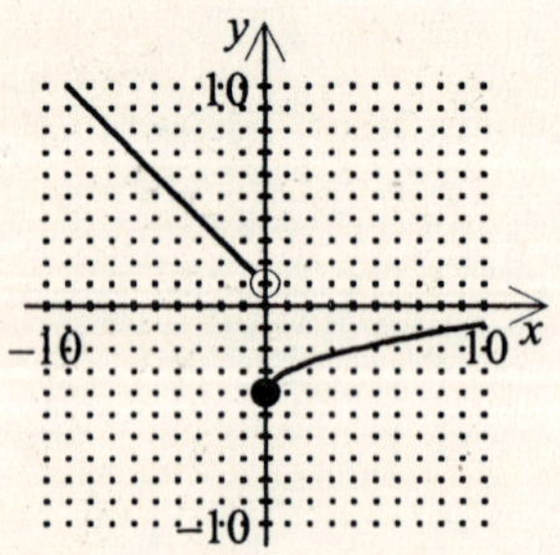

(B)

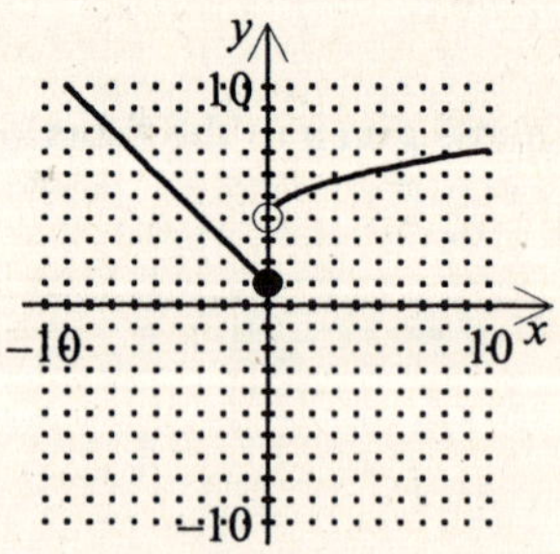

(C)

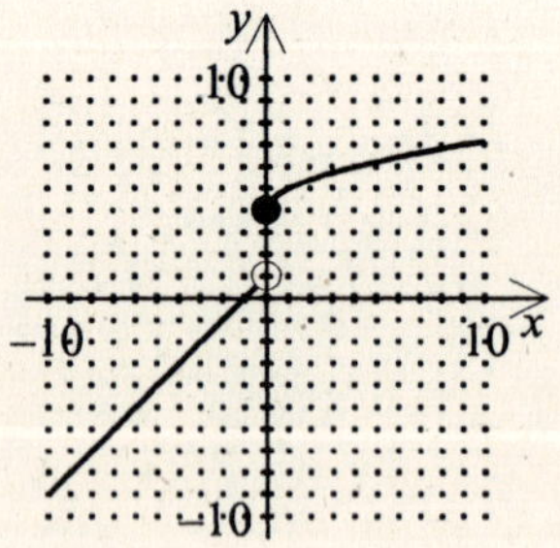

(D)

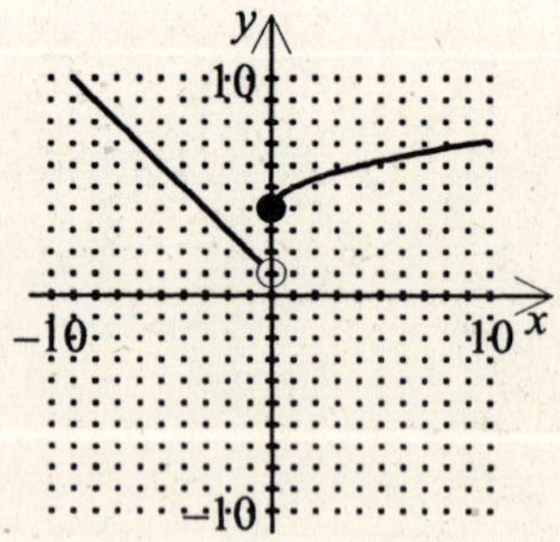

135. Graph the function.

$$f(x) = \begin{cases} 2, & x < 0 \\ x^2, & x \geq 0 \end{cases}$$

136. Sketch the graph of the function. (The symbol [[ ]] represents the greatest integer function.)

$$f(x) = [[3x - 4]]$$

Objective 4: Recognize graphs of common functions

137. Identify the common function type shown in the graph.

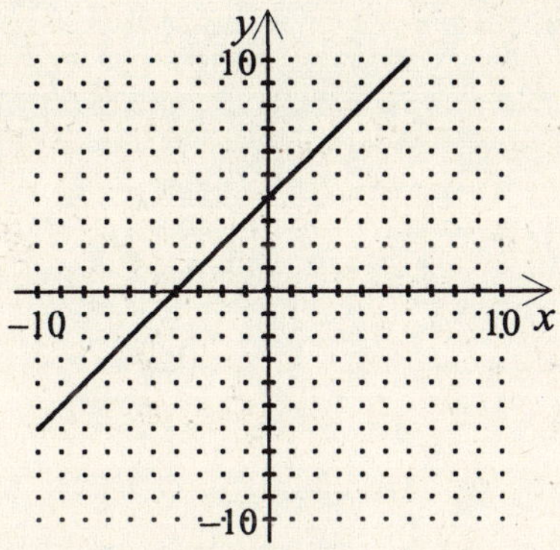

(A) Cubic Function
(B) Identity Function
(C) Quadratic Function
(D) Absolute Value Function

138. Find the equation for the common function shown in the graph.

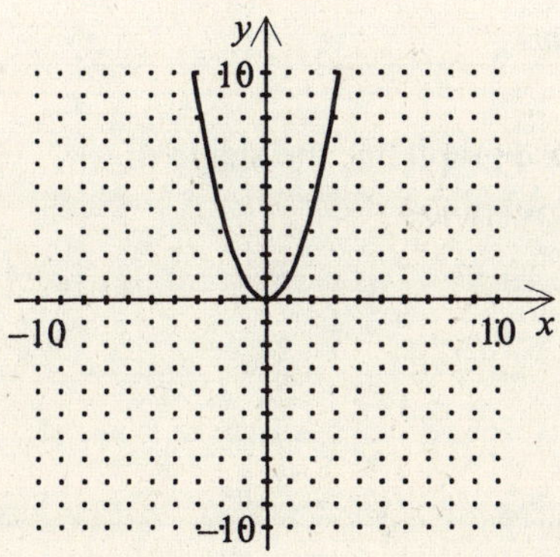

(A) $f(x) = x^2$
(B) $f(x) = c$
(C) $f(x) = x$
(D) $f(x) = |x|$

139. Identify the common function type shown in the graph.

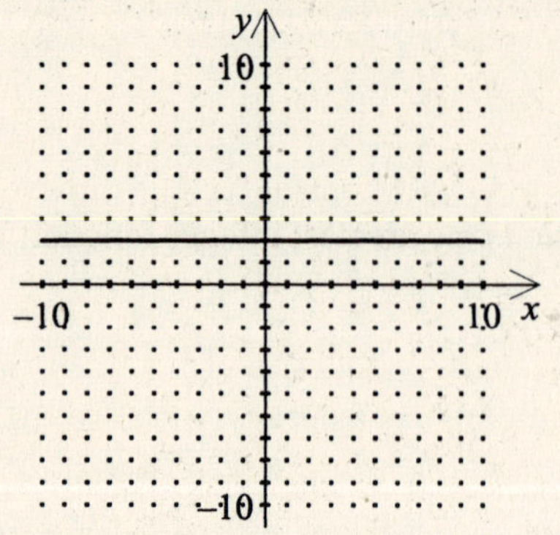

140. Identify the common function shown in the graph.

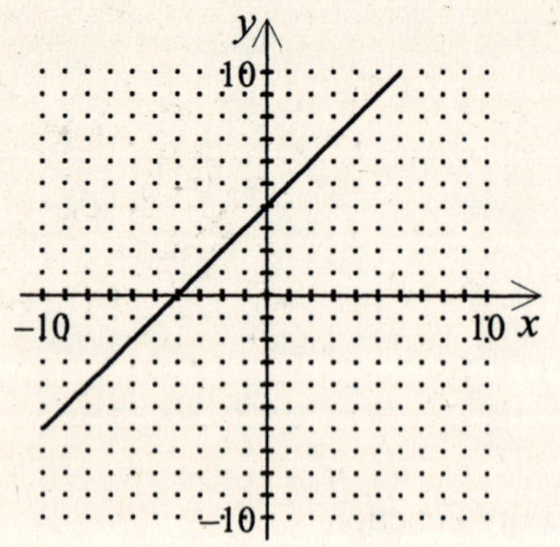

Section P.8: Shifting, Reflecting, and Stretching Graphs

Objective 1: Use vertical and horizontal shifts to sketch graphs of functions

141. Find an equation and identify the graph of the function whose graph has the shape of $f(x) = |x|$ but has moved 2 units to the right and 4 units downward.

(A) $f(x)=|x+2|+4$

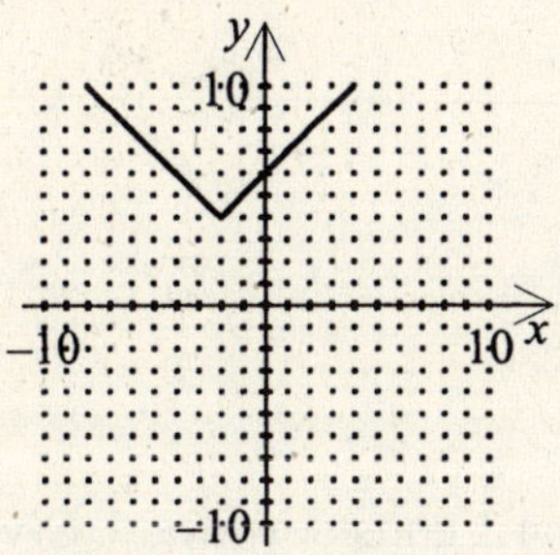

(B) $f(x)=|x-2|+4$

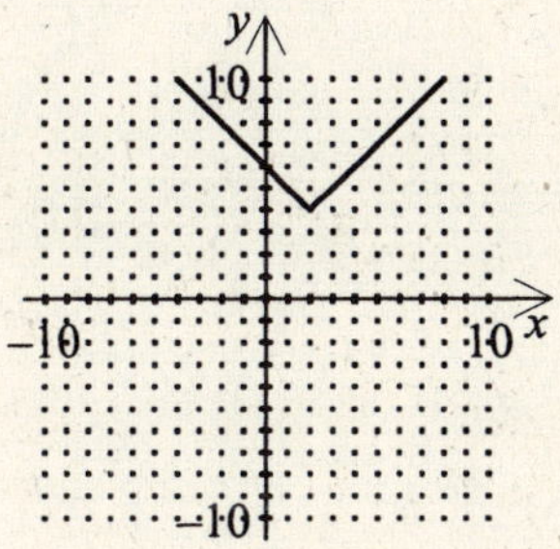

(C) $f(x)=|x-2|-4$

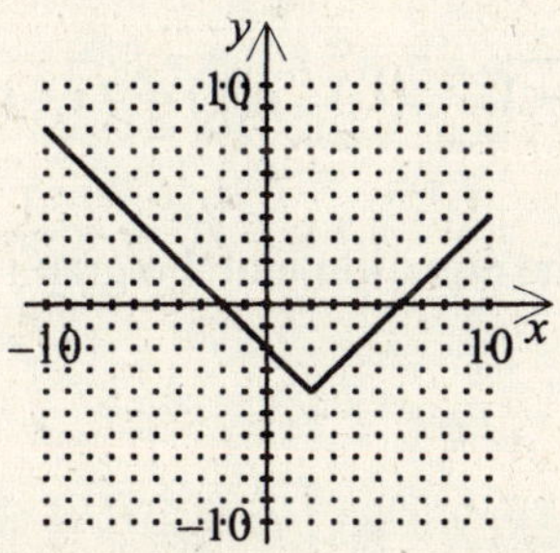

(D) $f(x)=|x+2|-4$

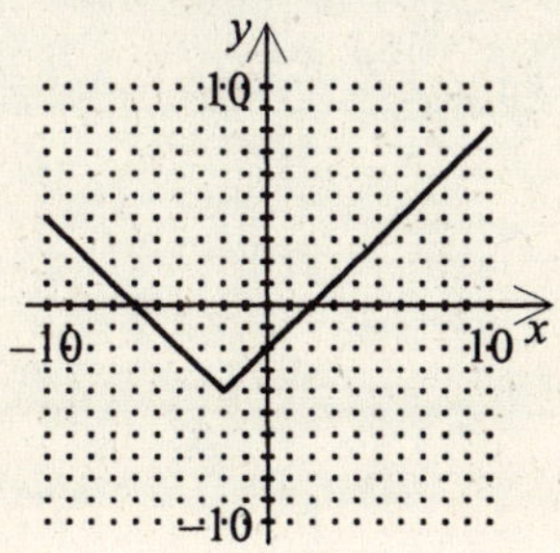

(141.)

142. Suppose $f(x)=x^3$ and $h(x)=(x+3)^3$. Determine which of the following statements is true.

(A) The graph *h* is the graph *f* shifted vertically down 3 units.

(B) The graph *h* is the graph *f* shifted horizontally left 3 units.

(C) The graph *h* is the graph *f* shifted vertically up 3 units.

(D) The graph *h* is the graph *f* shifted horizontally right 3 units.

143. Given $f(x)=x^2$, consider the translation 4 units to the left and 3 units downward. Give the translated function and its graph.

144. Sketch the graphs of the following functions.

(a) $f(x)=|x|$

(b) $h(x)=|x|+3$

(c) $k(x)=|x+3|$

Objective 2: Use reflections to sketch graphs of functions

145. Identify the function $g(x)$ that is the reflection of $f(x)=\sqrt{x}$ in the x-axis.

(A) $g(x)=-\sqrt{x}$ (B) $g(x)=\sqrt{-x}$ (C) $g(x)=-\sqrt{x-1}$ (D) $g(x)=\sqrt{x-1}$

146. The graph of the function $f(x)=|x|$ is shown below. Find the equation and the graph of the function $g(x)$ which is the reflection of $f(x)$ in the y-axis.

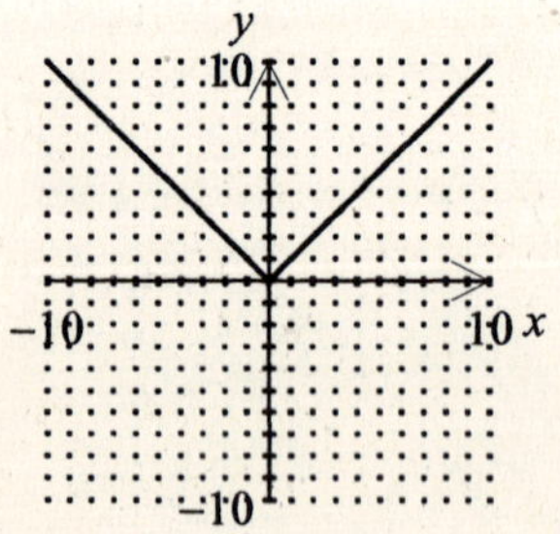

(A) $g(x)=|-x|$

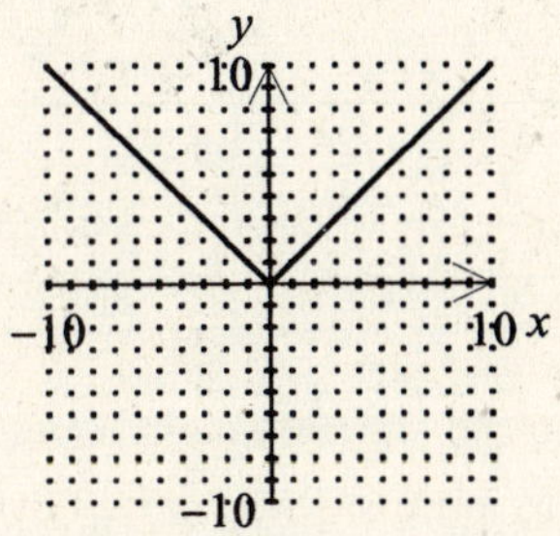

(B) $g(x)=|-x+1|$

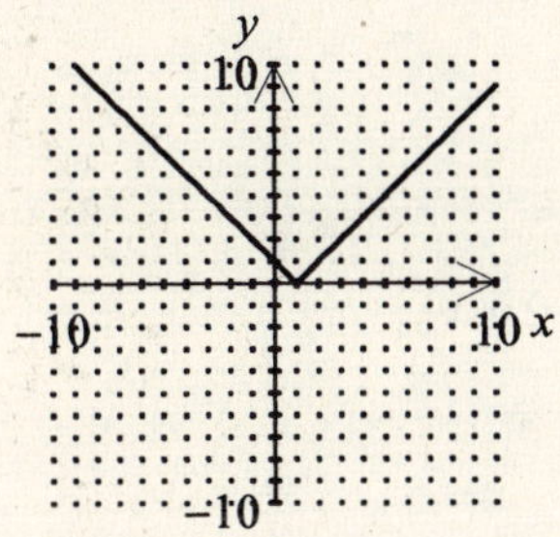

(C) $g(x)=-|x|$

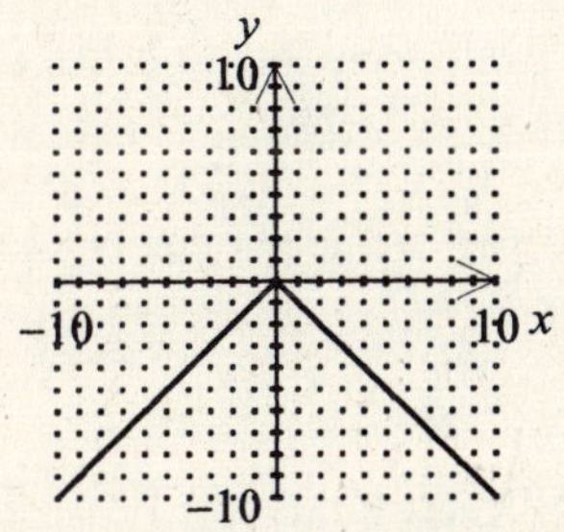

(D) $g(x)=-|x+1|$

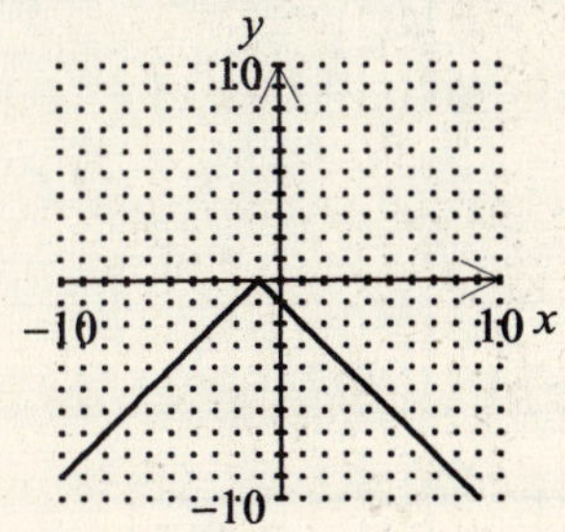

(146.)

147. Graph the pair of functions and identify the transformation. $f(x)=|x|+2,\ g(x)=-|x|-2$

148. Given below is the graph of a function $f(x)$. Graph $-f(x)$.

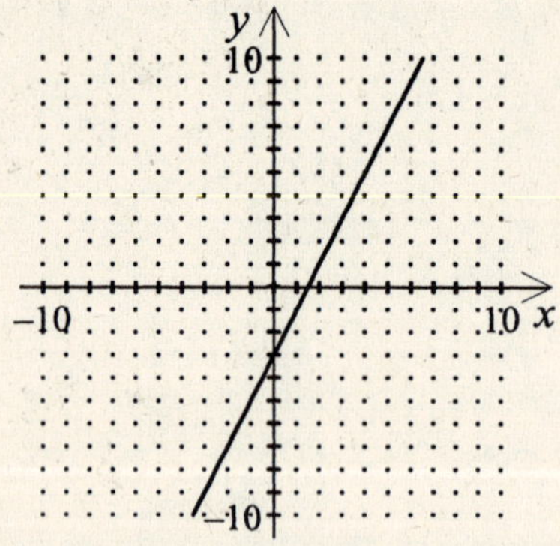

Objective 3: Use nonrigid transformations to sketch graphs of functions

149. The graph of the function $y(x)$ is given. Identify the graph of $2y(x)$.

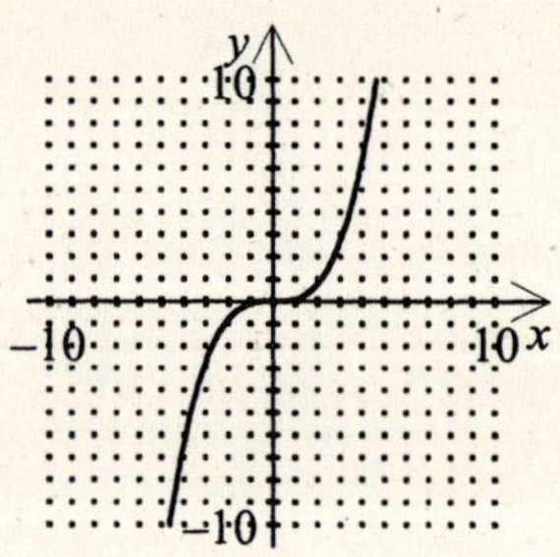

(A)

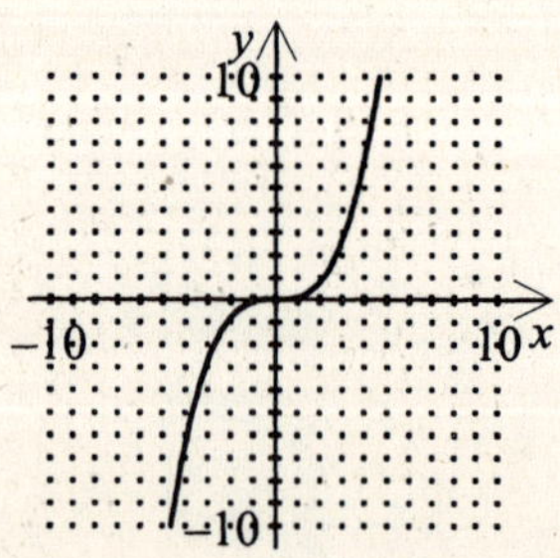

(B)

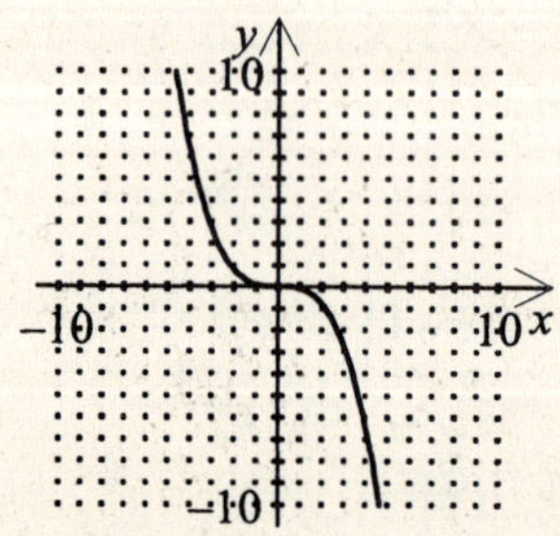

(C)

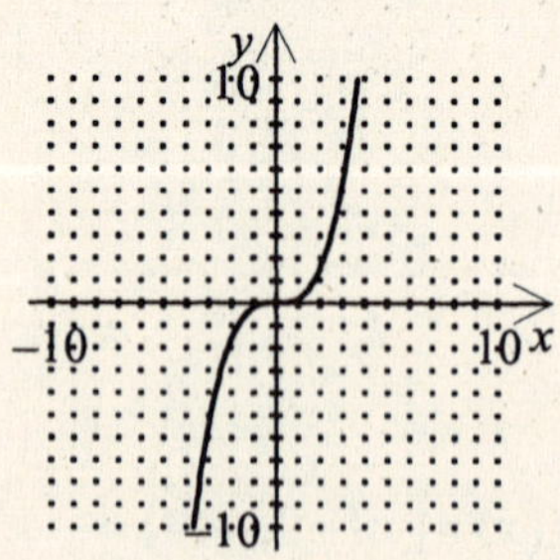

(D)

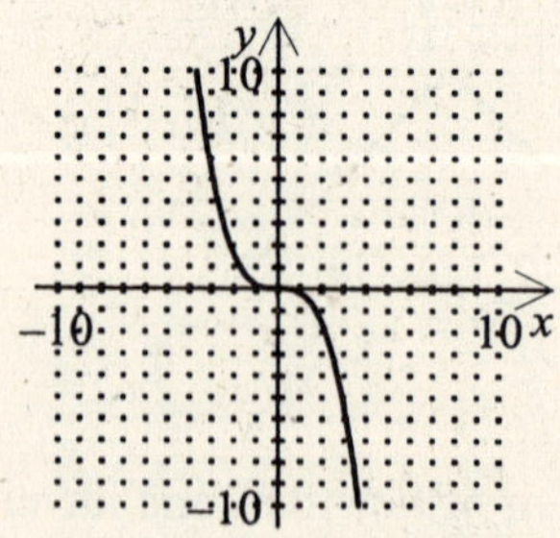

150. Find an equation and graph for the function that is described by the given characteristics.
The shape of $f(x)=|x|$, with a vertical stretch of 2.

(A) $g(x)=\frac{1}{2}|x|$

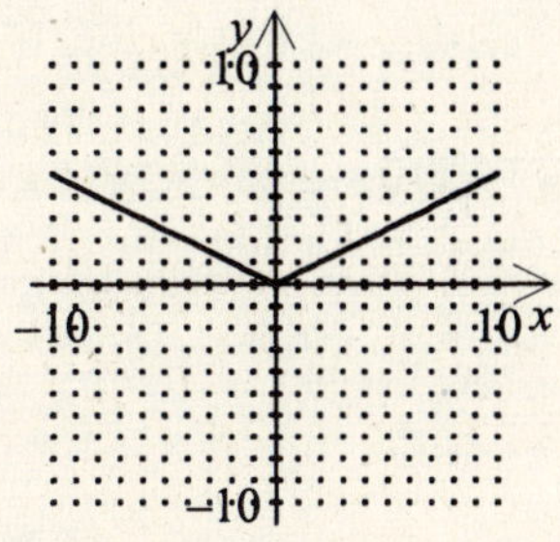

(B) $g(x)=2|x|$

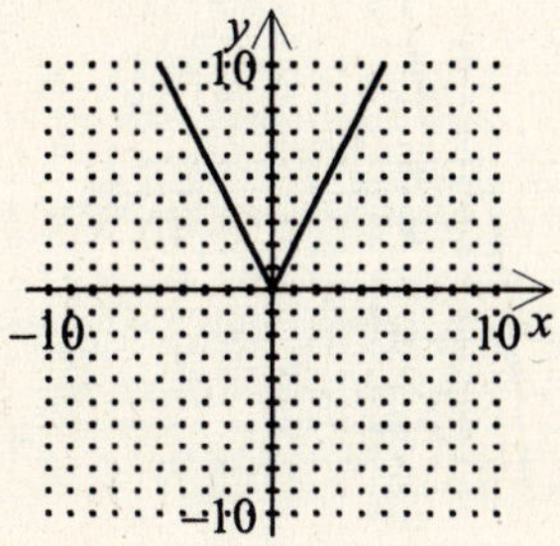

(C) $g(x)=-2|x|$

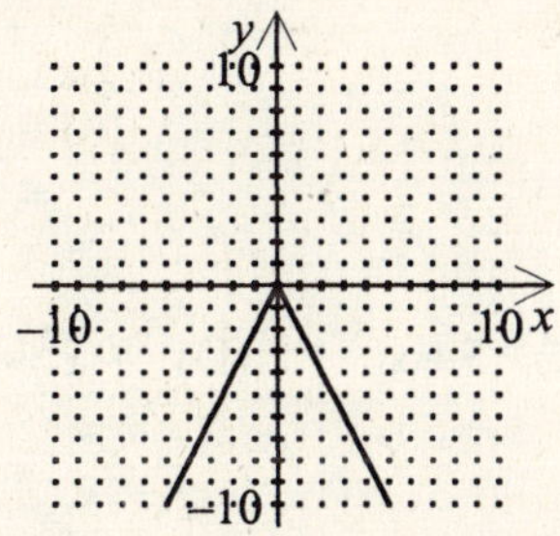

(D) $g(x)=-\frac{1}{2}|x|$

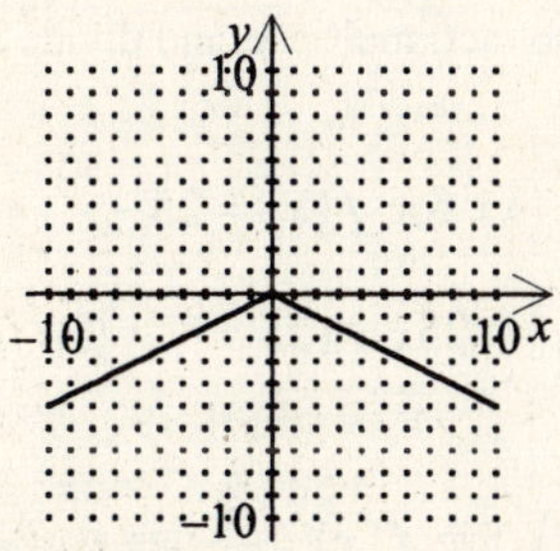

151. Write an equation for the function whose graph is shown.

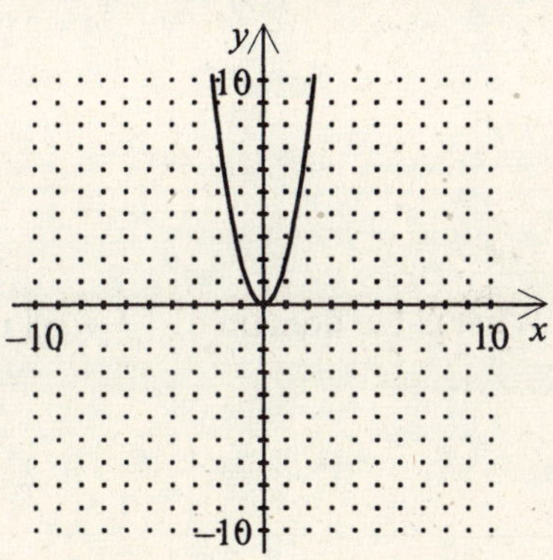

152. The graph of a common function is on the left and the graph of a transformation of that function is on the right. Find the equation of the transformed graph on the right.

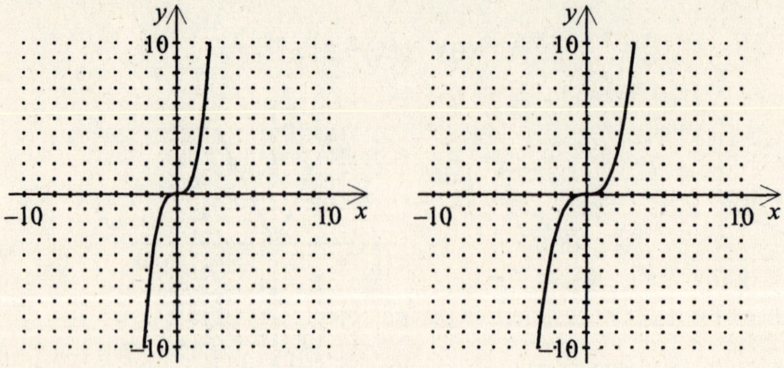

Section P.9: Combinations of Functions

Objective 1: Add, subtract, multiply, and divide functions

153. Find $(f+g)(x)$ for $f(x)=25-x^2$ and $g(x)=5-x$.

(A) $x^3-5x^2-25x+125$ (B) $-x^2-x+30$ (C) $5+x$ (D) $-x^2+x+20$

154. Find $\left(\dfrac{f}{g}\right)(x)$ for $f(x)=\dfrac{1}{x^2}$ and $g(x)=x$.

(A) x^3 (B) $\dfrac{1}{x^3}$ (C) $\dfrac{x^3-1}{x}$ (D) x^3-1

155. Evaluate $(f+g)(x)$ for $f(x)=4x+3$ and $g(x)=x-1$.

156. If $f(x)=6\sqrt{x}+4$, and $g(x)=-5\sqrt{x}+5$, find $(f+g)(x)$. What is the domain of $(f+g)(x)$?

Objective 2: Find the compositions of one function with another function

157. If $f(x)=|5x|$ and $g(x)=-2x$, find $(g\circ f)(x)$.

(A) $|5x|+2x$ (B) $10|x|$ (C) $-10|x|$ (D) $-2x+|5x|$

158. If $f(x)=7x+8$ and $g(x)=4x-2$, find $(g\circ f)(-5)$.

(A) −146 (B) −109 (C) −147 (D) −110

159. If $f(x)=x-5$ and $g(x)=\sqrt{x+1}$, find $(f\circ g)(x)$.

160. If $f(x)=4-7x$ and $g(x)=x^2+2$, find (a) $(f\circ g)(x)$ and (b) $(g\circ f)(x)$.

Objective 3: Use combinations of functions to model and solve real-life problems

161. The number of pounds of raspberries a cannery can process and the processing cost are
$P(h)=300h$ and $C(n)=0.4n+500$
where $P(h)$ is the number of pounds of raspberries that can be processed in h hours and $C(n)$ is the cost of processing n pounds of raspberries. Use composition functions to find the cost of operating the cannery 76 hours.

(A) \$23,000 (B) \$9920 (C) \$9620 (D) \$9320

162. The selling price of x number of a certain stereo can be modeled by the function $R(x)=170x$.
The total cost of making x stereos is
$C(x)=77x-0.02x^2$.
What is the percent markup for 40 stereos?

(A) 375% (B) 123% (C) 223% (D) 22%

163. The production in board feet and the cost of manufacturing lumber at a sawmill are
$P(h)=400h$ and $C(n)=0.25n+10{,}000$
where $P(h)$ is the number of board feet that can be produced in h hours and $C(n)$ is the cost of producing n board feet. Find $(C\circ P)(h)$ and interpret its meaning.

164. Spheres are being packed into a square box (see figure).
(a) Express the radius r of each sphere as a function of the length x of the sides of the square.
(b) Express the volume V of a sphere as a function of the radius r.
(c) Find $(V\circ r)(x)$.
(d) Find and interpret $(V\circ r)(3)$.

Section P.10: Inverse Functions

Objective 1: Find inverse functions informally and verify that two functions are inverse functions of each other

165. Find the inverse of the function.

$f(x) = 2x - \frac{1}{2}$

(A) $f^{-1}(x) = \frac{1}{4}x - \frac{1}{4}$ (B) $f^{-1}(x) = \frac{1}{2}x + \frac{1}{4}$ (C) $f^{-1}(x) = \frac{1}{2}x + 1$ (D) $f^{-1}(x) = 2x - 1$

166. Are the following functions inverses? If not, rewrite the second function so that it is an inverse of the first.

$f(x) = 3x - 4$

$g(x) = \frac{1}{3}x - \frac{4}{3}$

(A) Yes (B) No; $g(x) = 3x + \frac{4}{3}$ (C) No; $g(x) = \frac{1}{3}x + \frac{4}{3}$ (D) No; $g(x) = x - \frac{4}{3}$

Find the inverse of *f* informally. Verify that $f(f^{-1}(x)) = x$ and $f^{-1}(f(x)) = x$.

167. $f(x) = 6x$

168. $f(x) = \sqrt[3]{x}$

Objective 2: Use graphs of functions to determine whether functions have inverse functions

169. Use the Horizontal and Vertical Line Tests to determine which of the following is a function that has an inverse.

(i)

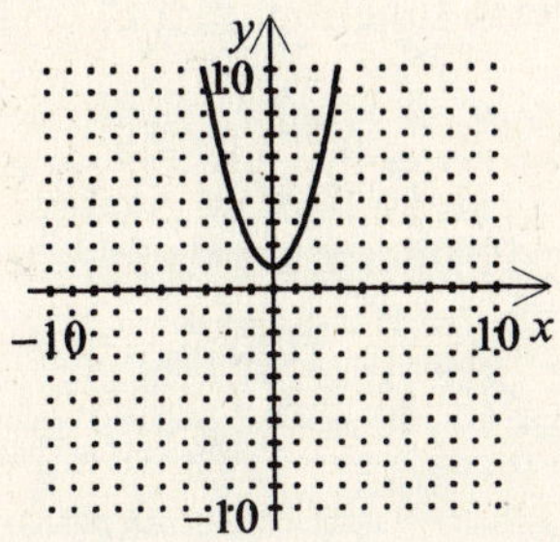

(ii)

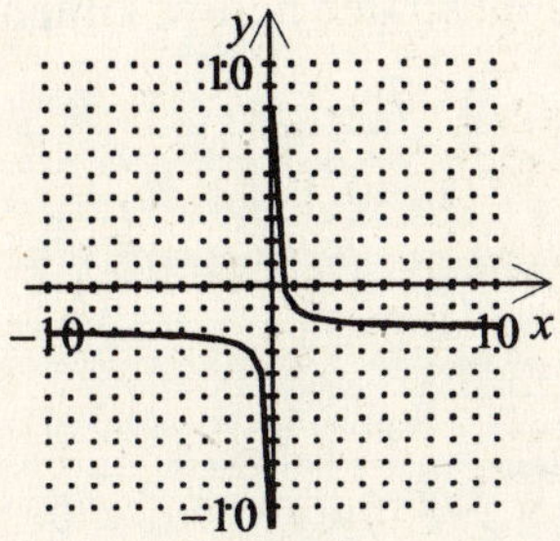

(iii)

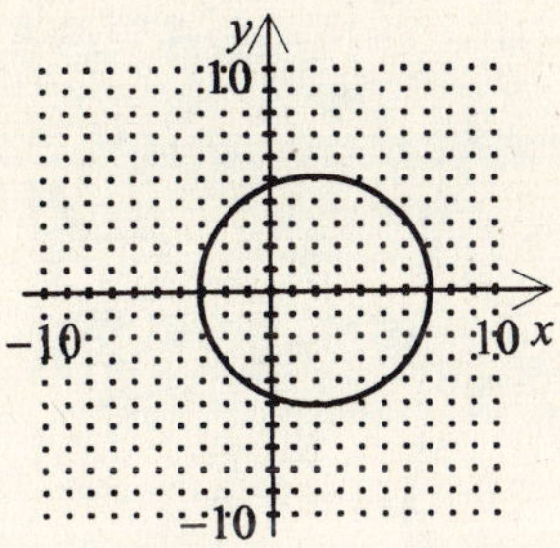

(iv)

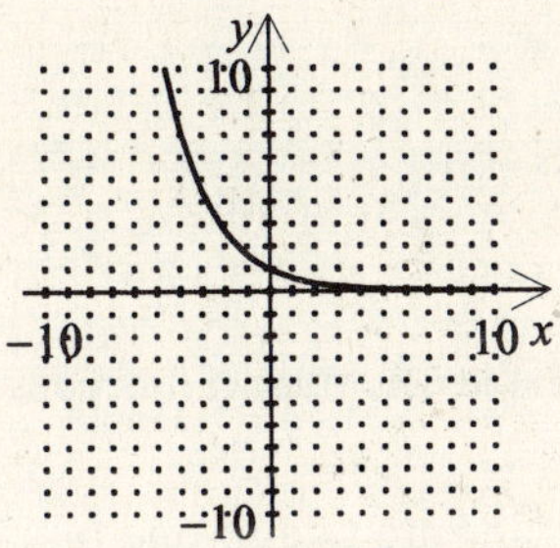

(A) ii and iv only (B) iv only (C) i and iv only (D) i, ii, and iv only

170. Let $f(x) = 2x - 2$. Identify the graph of f and f^{-1}.

(A)

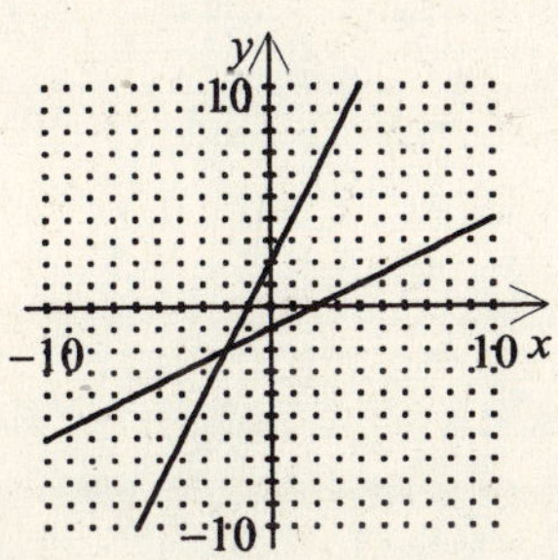

(B)

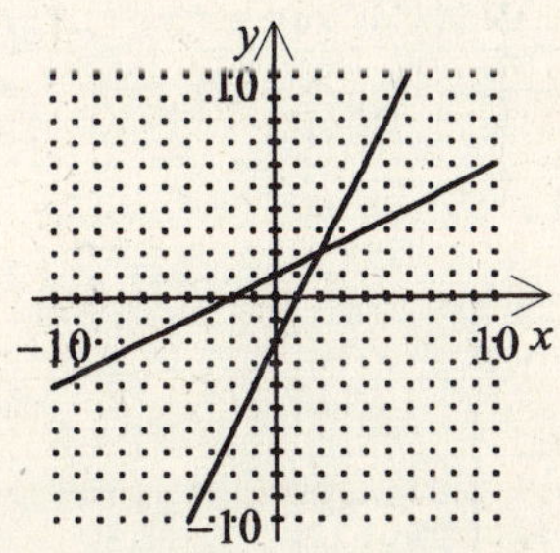

(C)

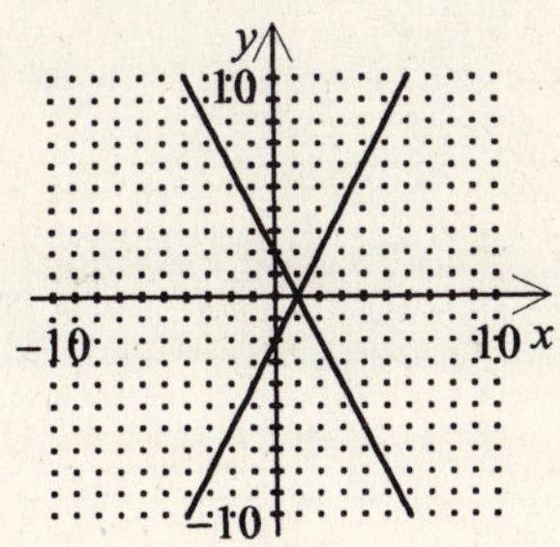

(D)

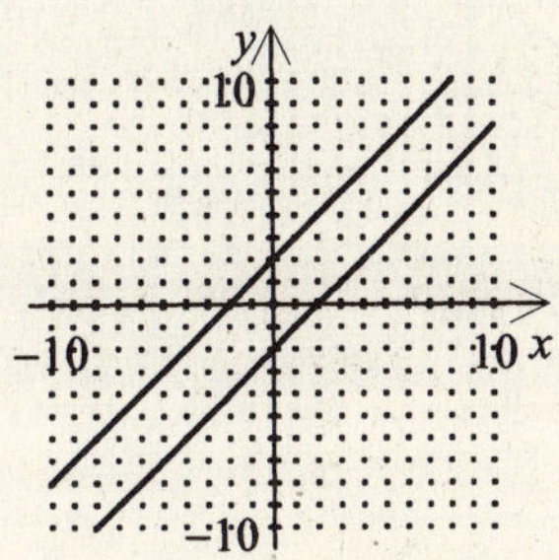

171. Use a graphing utility to graph the function $f(x) = |x+2|+2$ and use the Horizontal Line Test to determine whether the function has an inverse.

172. The graph of a function $f(x)$ is illustrated below. Graph the inverse function $f^{-1}(x)$.

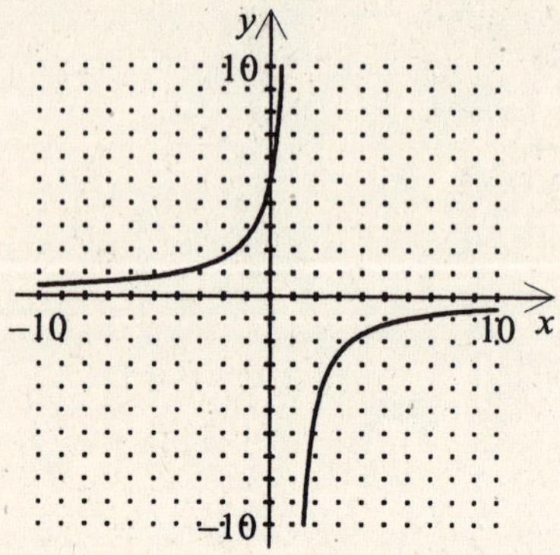

Objective 3: Use the Horizontal Line Test to determine if functions are one-to-one

173. Determine which of the following are one-to-one functions.

i.

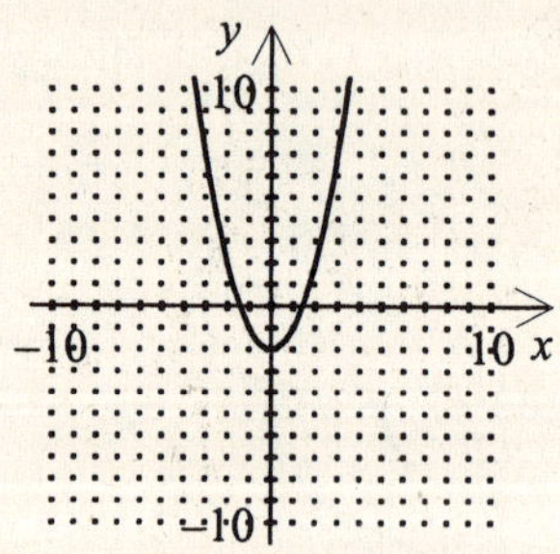

ii.

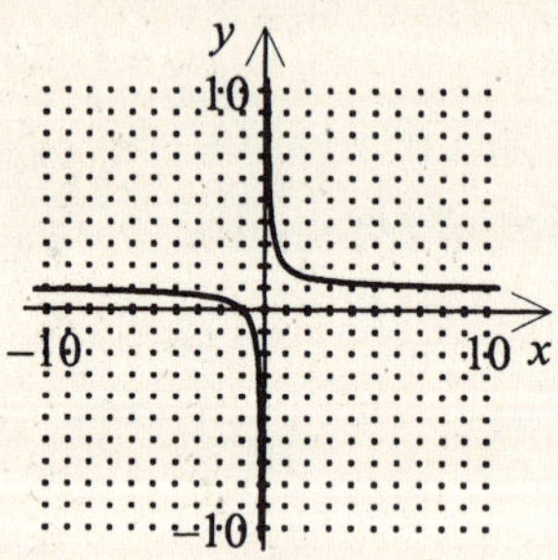

iii.

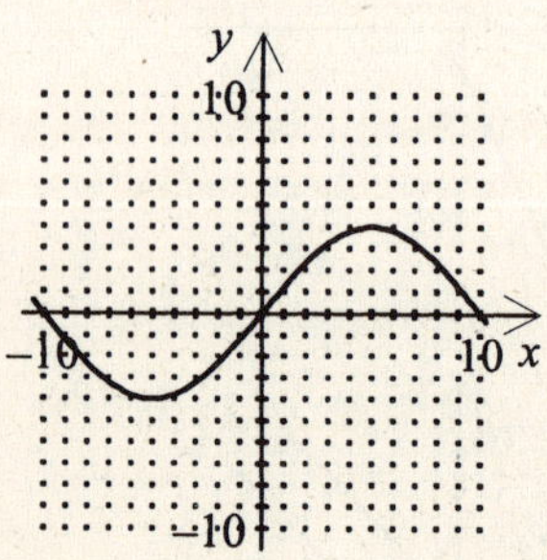

iv.

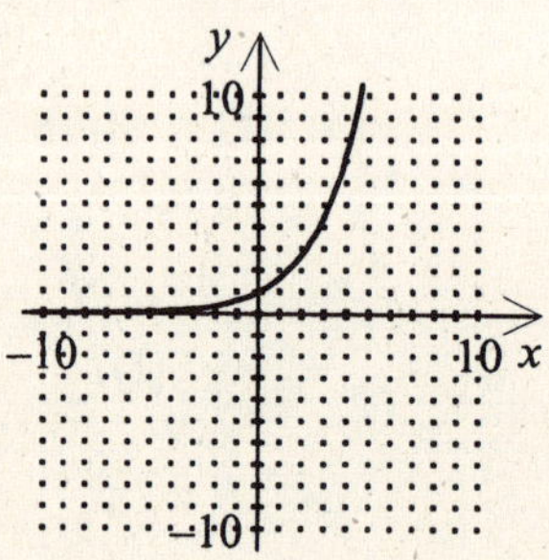

(A) ii and iv only (B) iv only (C) i, ii, and iv only (D) i and iv only

174. Which of the following is a one-to-one function?

(A) $\{(0,\ 9),\ (-9,\ 7),\ (0,\ 3),\ (7,\ -9)\}$

(B) $\{(0,\ 9),\ (-9,\ -8),\ (-8,\ -9),\ (-6,\ -6)\}$

(C) $\{(0,\ 9),\ (-9,\ 7),\ (-8,\ -8),\ (-9,\ -2)\}$

(D) $\{(0,\ 9),\ (-9,\ 7),\ (-8,\ 3),\ (-6,\ 9)\}$

175. Is $3x + 5y = 6$ a one-to-one function?

176. Which of the following are one-to-one functions?

i.

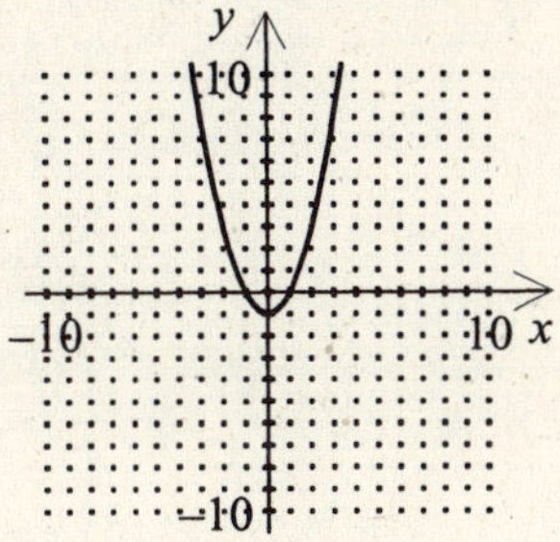

ii.

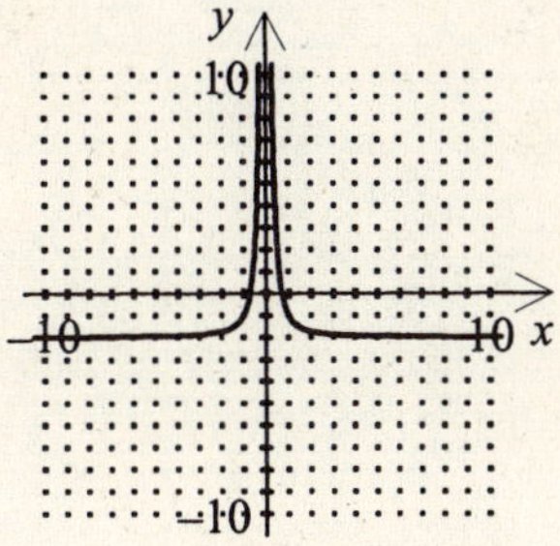

iii.

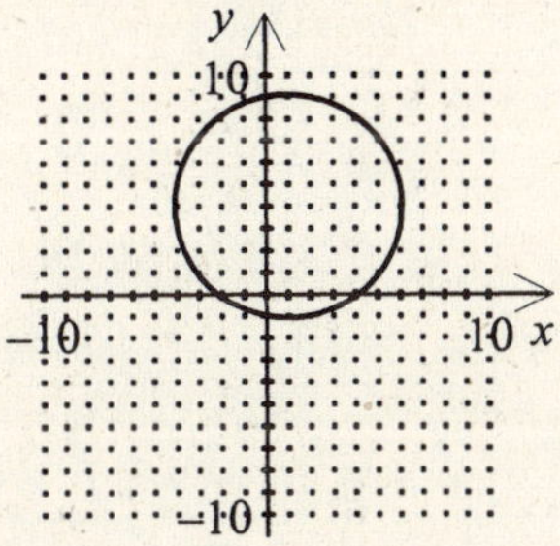

iv.

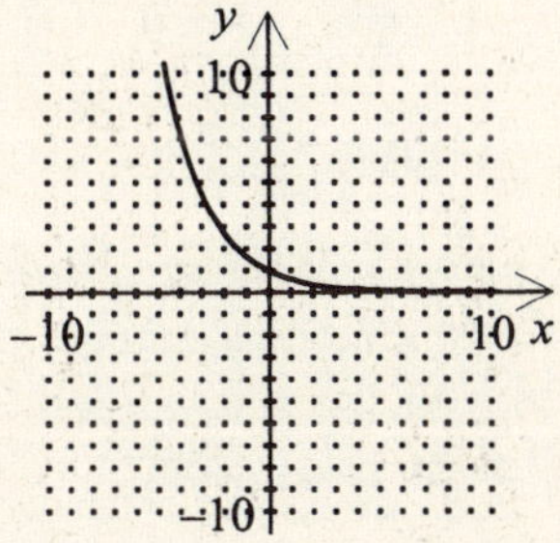

Objective 4: Find inverse functions algebraically

Find the inverse of the function.

177. $f(x) = \dfrac{-4-2x}{-1-5x}$

(A) $f^{-1}(x) = \dfrac{x-4}{-5x+2}$

(B) $f^{-1}(x) = \dfrac{-2-4x}{-5-x}$

(C) $f^{-1}(x) = \dfrac{-5x-1}{-2x-4}$

(D) $f^{-1}(x) = \dfrac{-5-x}{-2-4x}$

Find the inverse of the function.

178. $f(x)=4\sqrt[3]{3x-2}+10$

(A) $f^{-1}(x)=\dfrac{x^3+30x^2+300x-872}{192}$

(B) $f^{-1}(x)=\dfrac{x^3-30x^2+300x-872}{192}$

(C) $f^{-1}(x)=\dfrac{x^3-30x^2+300x-1128}{192}$

(D) $f^{-1}(x)=\dfrac{x^3+30x^2+300x-1128}{192}$

179. $f(x)=\dfrac{x+7}{6}$

180. $f(x)=\sqrt{5x+1}-1$

Answer Key for Chapter P Prerequisites

Section P.1: Review of Real Numbers and Their Properties

Objective 1: Represent and classify real numbers

[1] (A)

[2] (C)

[3] $25,\ 3.56,\ 0,\ -7.3333,\ \frac{3}{5},\ -8$

[4] $0.\overline{63}$

Objective 2: Order real numbers and use inequalities

[5] (B)

[6] (C)

[7] $-\frac{7}{16} < -\frac{2}{7}$

[8] $x > -5$ denotes the set of all real numbers greater than -5.

The interval is unbounded.

Objective 3: Find the absolute values of real numbers and find the distance between two real numbers

[9] (D)

[10] (C)

[11] $2,\ |-7|,\ 9,\ |-16|$

[12] -6

Objective 4: Evaluate algebraic expressions

[13] (C)

[14] (A)

[15] –7

[16] 74

Objective 5: Use the basic rules and properties of algebra

[17] (D)

[18] (B)

[19] Answers may vary. Sample answer: $9+0=9$

[20] $\frac{27}{25}$

Section P.2: Solving Equations

Objective 1: Identify different types of equations

[21] (C)

[22] (C)

[23] Identity

[24] Identity

Objective 2: Solve linear equations in one variable and equations that lead to linear equations

[25] (D)

[26] (C)

[27] $-\frac{13}{2}$

[28] 21

Objective 3: Solve quadratic equations by factoring, extracting square roots, completing the square, and using the Quadratic Formula

[29] (B)

[30] (D)

[31] $-5, -3$

[32] $\frac{4 \pm \sqrt{10}}{2}$

Objective 4: Solve polynomial equations of degree three or greater

[33] (B)

[34] (C)

[35] $\pm 1, 5$

[36] $-\frac{7}{2}$

Objective 5: Solve equations involving radicals

[37] (C)

[38] (B)

[39] -9

[40] 2

Objective 6: Solve equations involving absolute values

[41] (D)

[42] (C)

[43] −2, 3

[44] $1, -\frac{5}{3}$

Section P.3: The Cartesian Plane and Graphs of Equations

Objective 1: Plot points in the Cartesian plane

[45] (A)

[46] (C)

[47] (3, 3); Quadrant I

[48] *B*

Objective 2: Use the Distance Formula to find the distance between two points

[49] (A)

[50] (A)

[51] 14

[52] 1.2

Objective 3: Use the Midpoint Formula to find the midpoint of a line segment

[53] (D)

[54] (C)

[55] $(-2, 0)$

[56] $(1, 1)$; $d \approx 2.24$

Objective 4: Sketch graphs of equations

[57] (C)

[58] (C)

[59]

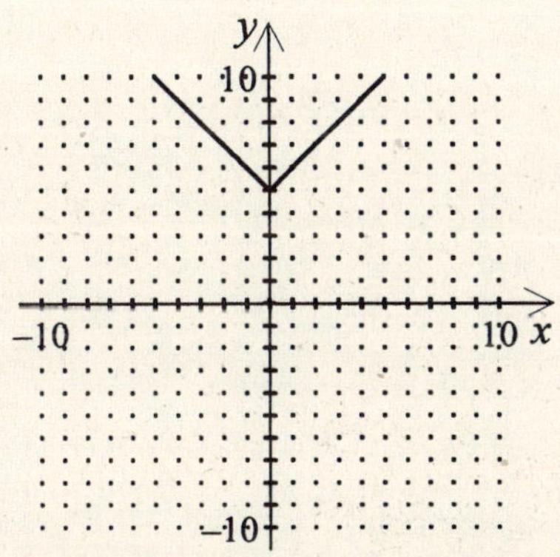

x	-2	-1	0	1	2
y	-14	-9	-4	1	6

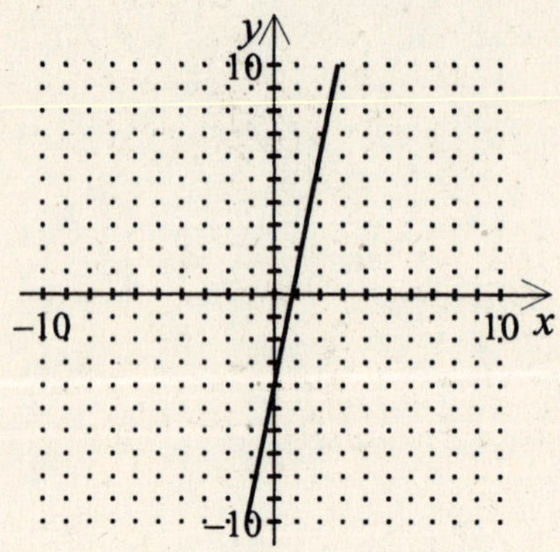

[60]

Objective 5: Find x- and y-intercepts of graphs

[61] (D)

[62] (C)

x-intercept: none
[63] y-intercept: $(0,\ 6)$

x-intercept: none
[64] y-intercept: $(0,\ 6)$

Objective 6: Use symmetry to sketch graphs of equations

[65] (D)

[66] (D)

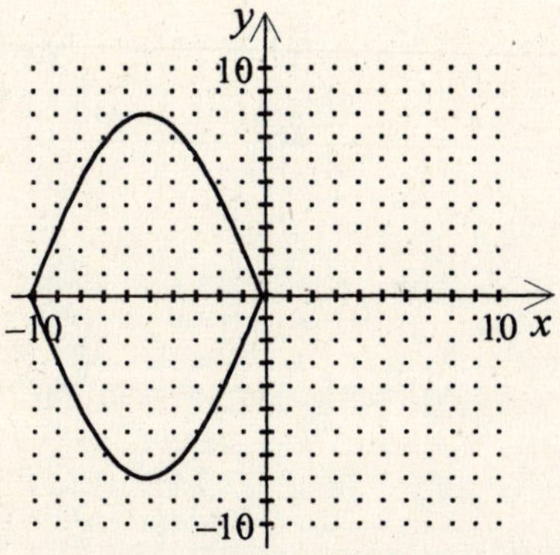

[67] ______________________

[68] The graph is symmetric with respect to the y-axis.

Objective 7: Find equations and sketch graphs of circles

[69] (A)

[70] (A)

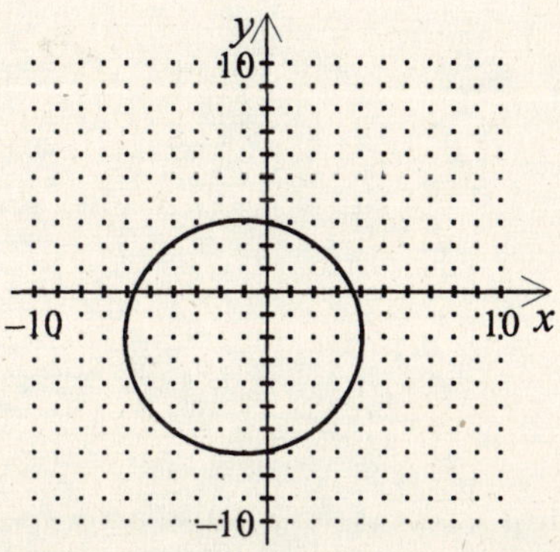

[71] ______________________

[72] $(x+1)^2 + (y-2)^2 = 61$

Section P.4: Linear Equations in Two Variables

Objective 1: Use slope to graph linear equations in two variables

[73] (C)

[74] (A)

[75] Answers may vary. Sample answer: $(6, 0)$, $(7, 1)$, $(8, 2)$

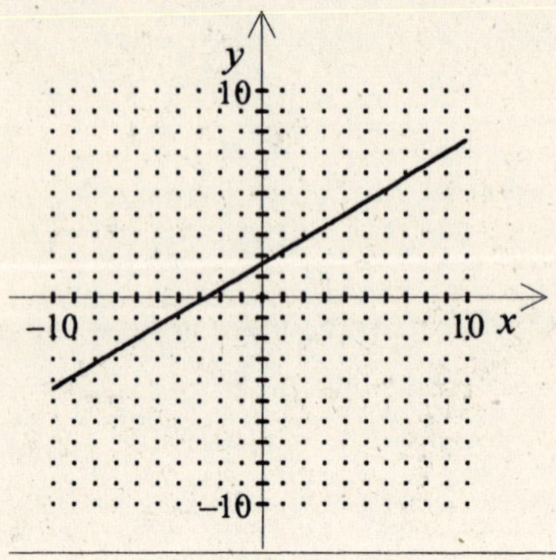

[76]

Objective 2: Find slopes of lines

[77] (B)

[78] (B)

[79] $\frac{5}{3}$

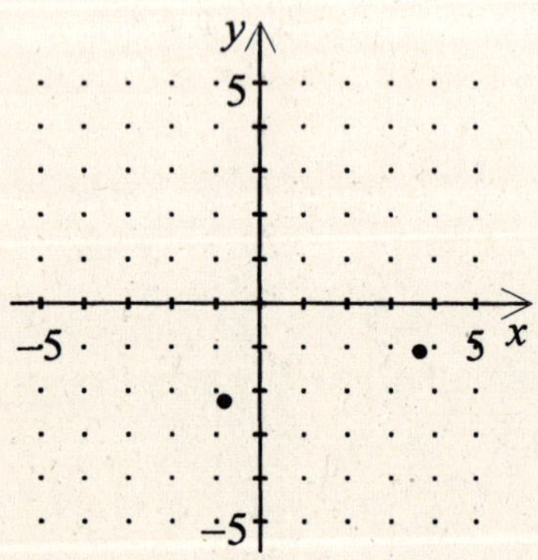

[80] $\frac{205}{804}$

Objective 3: Write linear equations in two variables

[81] (C)

[82] (A)

[83] $y=9$

[84] $-\frac{x}{4}+\frac{y}{6}=1$

Objective 4: Use slope to identify parallel and perpendicular lines

[85] (C)

[86] (C)

[87] $y=-\frac{7}{4}x+\frac{9}{2}$

[88] Parallel

Objective 5: Use linear equations in two variables to model and solve real-life problems

[89] (B)

[90] (B)

[91] (a) $y=6x+6$
(b) The y-intercept represents the average weight of a kitten at birth. The slope indicates the average weekly weight gain for these kittens during their first 5 weeks.

[92] $y=-\frac{2}{5}x+13$

Section P.5: Functions

Objective 1: Determine whether relations between two variables are functions

[93] (D)

[94] (D)

[95] Yes, there is only one y value for each x value.

[96] No

Objective 2: Use function notation and evaluate functions

[97] (D)

[98] (D)

[99] (a) −1 (b) 8 (c) 7 (d) 0.16

[100] $f(-27) = -12$
$f(54) = 51$

Objective 3: Find domains of functions

[101] (C)

[102] (A)

[103] $x \geq -10,\ x \neq 2,\ 4$

[104] $x \geq 0$

Objective 4: Use functions to model and solve real-life problems

[105] (C)

[106] (C)

[107] 462,000

[108] $H(d) = 0.8d + 14$

Section P.6: Analyzing Graphs of Functions

Objective 1: Use the Vertical Line Test for functions

[109] (A)

[110] (B)

[111] Function

[112] Not a function

Objective 2: Find the zeros of functions

[113] (D)

[114] (C)

[115] 4

[116] $-1, 0, 6$

Objective 3: Determine intervals on which functions are increasing or decreasing

[117] (B)

[118] (B)

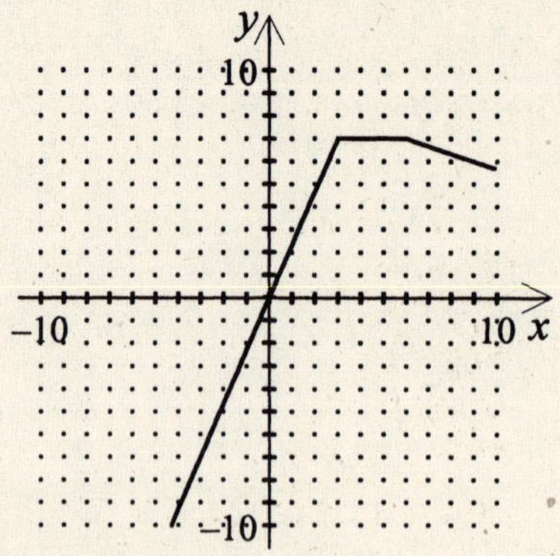

Increasing on $(-\infty, 3]$

Constant on $(3, 6]$

[119] Decreasing on $(6, \infty)$

(a)

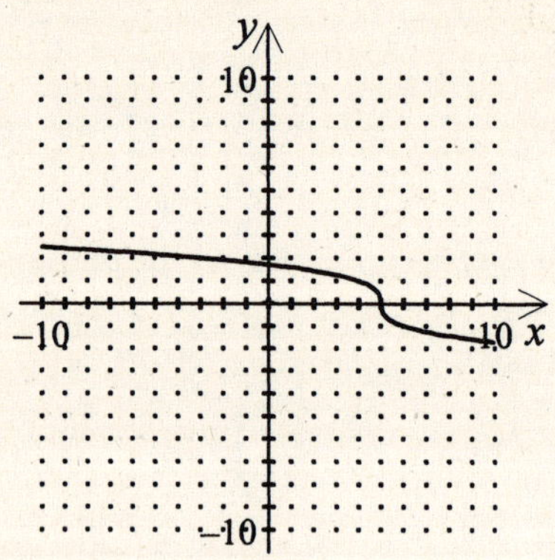

Decreasing on $(-\infty, \infty)$

[120] (b) Sample:

x	3	4	6	7
$f(x)$	1.26	1	-1	-1.26

Objective 4: Identify even and odd functions

[121] (C)

[122] (A)

[123] Neither

[124] Even

Section P.7: A Library of Functions

Objective 1: Identify and graph linear and squaring functions

[125] (D)

[126] (B)

[127] 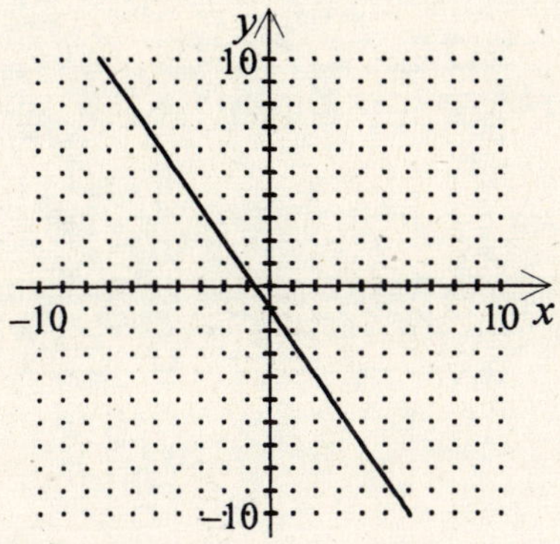

[128] $f(x) = x - 2$

Objective 2: Identify and graph cubic, square root, and reciprocal functions

[129] (D)

[130] (D)

[131] (0, 0)

[132] reciprocal function

Objective 3: Identify and graph step and other piecewise-defined functions

[133] (A)

[134] (D)

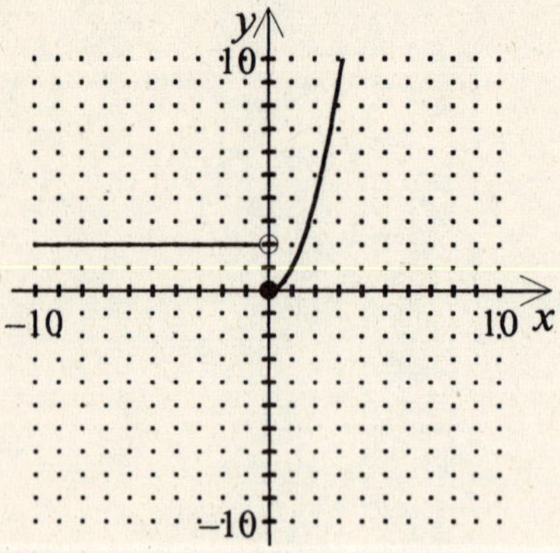

[135] ______________________

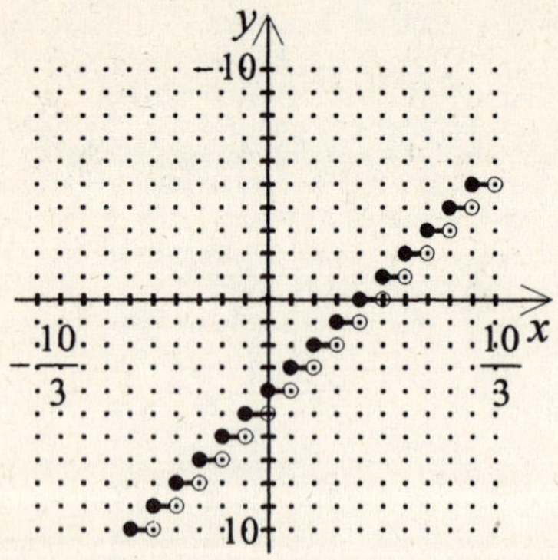

[136] ______________________

Objective 4: Recognize graphs of common functions

[137] (B)

[138] (A)

[139] Constant Function

[140] $f(x) = x$

Section P.8: Shifting, Reflecting, and Stretching Graphs

Objective 1: Use vertical and horizontal shifts to sketch graphs of functions

[141] (C)

[142] (B)

$f(x) = (x+4)^2 - 3$

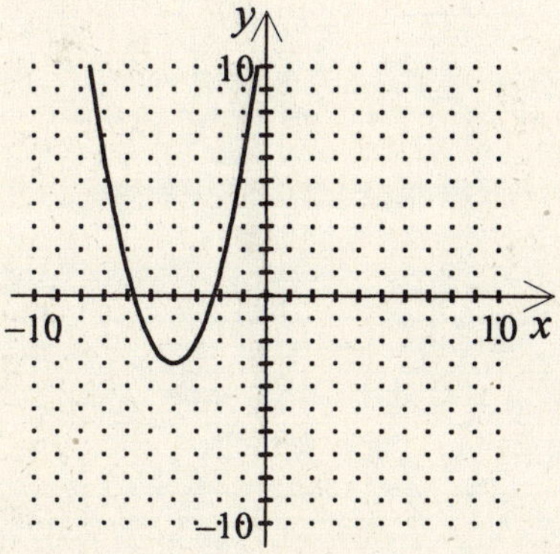

[143]

(a)

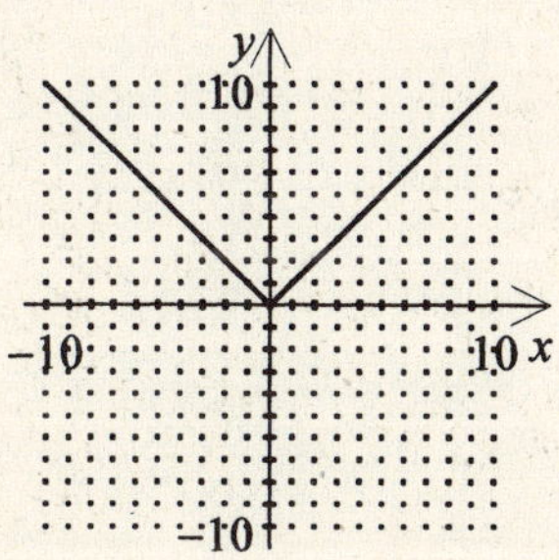

(b)

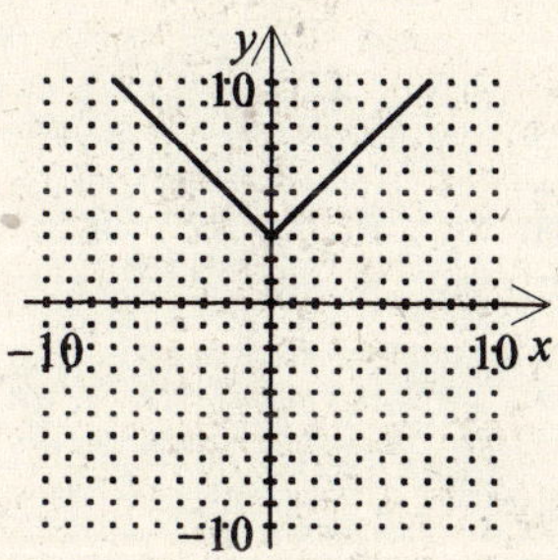

(c)

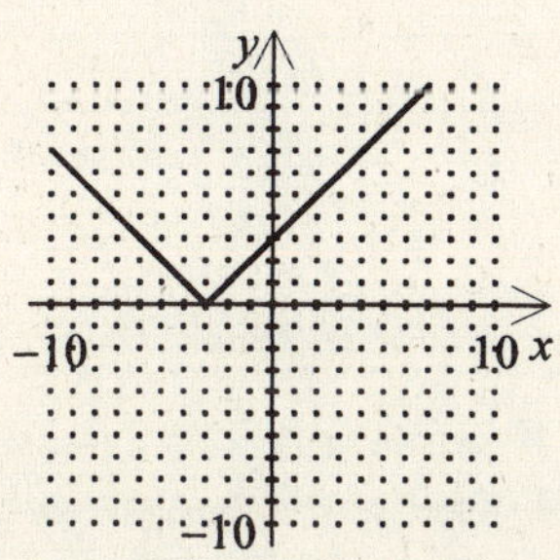

[144]

Objective 2: Use reflections to sketch graphs of functions

[145] (A)

[146] (A)

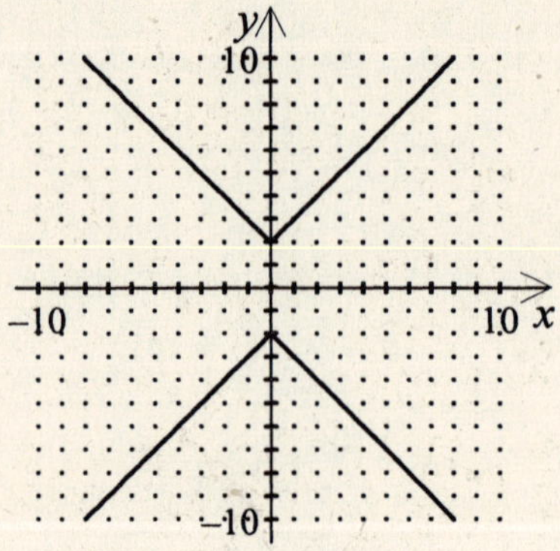

[147] Reflection in the x-axis

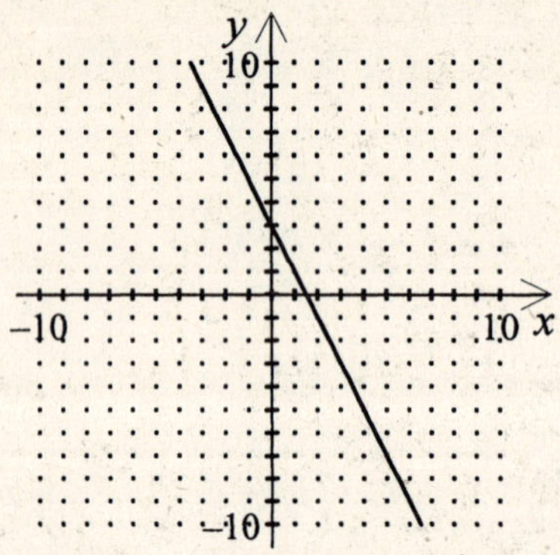

[148]

Objective 3: Use nonrigid transformations to sketch graphs of functions

[149] (C)

[150] (B)

[151] $f(x) = 2x^2$

[152] $f(x) = \frac{1}{3}x^3$

Section P.9: Combinations of Functions

Objective 1: Add, subtract, multiply, and divide functions

[153] (B)

[154] (B)

[155] $5x+2$

[156] $(f+g)(x)=\sqrt{x}+9,\ x \geq 0$

Objective 2: Find the compositions of one function with another function

[157] (C)

[158] (D)

[159] $\sqrt{x+1}-5$

[160] (a) $-7x^2-10$
(b) $49x^2-56x+18$

Objective 3: Use combinations of functions to model and solve real-life problems

[161] (C)

[162] (C)

[163] $(C \circ P)(h)=100h+10{,}000$; the cost of operating h hours.

(a) $r(x) = \frac{x}{6}$

(b) $V(r) = \frac{4}{3}\pi r^3$

(c) $(V \circ r)(x) = \frac{x^3}{162}\pi$

[164] (d) $(V \circ r)(3) = 0.524$. The volume of one sphere when the length of each side of the square is 3.

Section P.10: Inverse Functions

[165] (B)

[166] (C)

[167] $f^{-1}(x) = \frac{1}{6}x$

[168] $f^{-1}(x) = x^3$

Objective 2: Use graphs of functions to determine whether functions have inverse functions

[169] (A)

[170] (B)

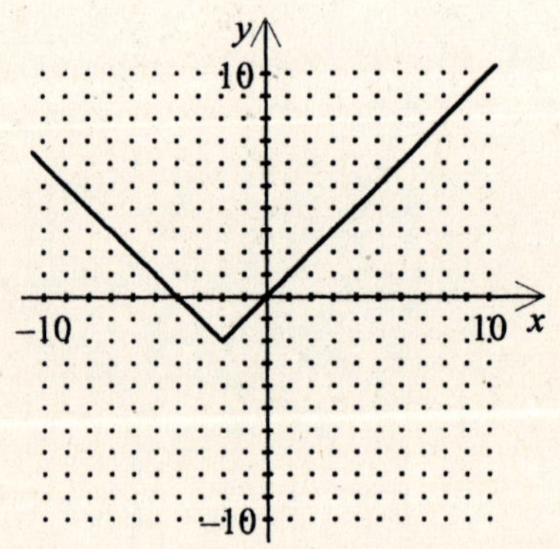

[171] No

[172] 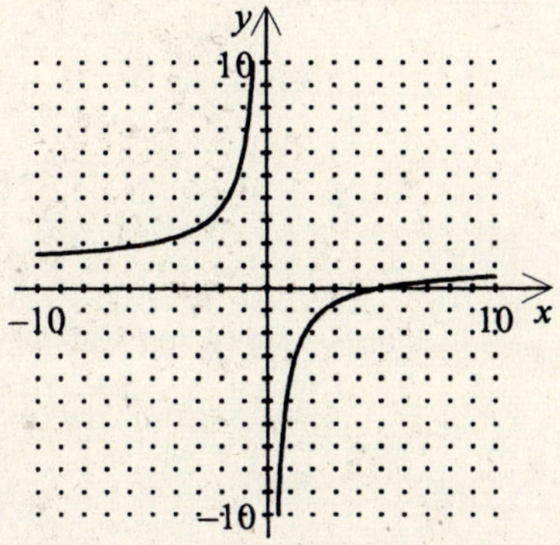

Objective 3: Use the Horizontal Line Test to determine if functions are one-to-one

[173] (A)

[174] (B)

[175] Yes

[176] iv only

Objective 4: Find inverse functions algebraically

[177] (A)

[178] (B)

[179] $f^{-1}(x) = 6x - 7$

[180] $f^{-1}(x) = \dfrac{x^2 + 2x}{5}, \ x \geq 0$

Chapter 1 Trigonometry

Section 1.1: Radian and Degree Measure

Objective 1: Describe angles

1. Identify the angle that is in standard position.

(A)

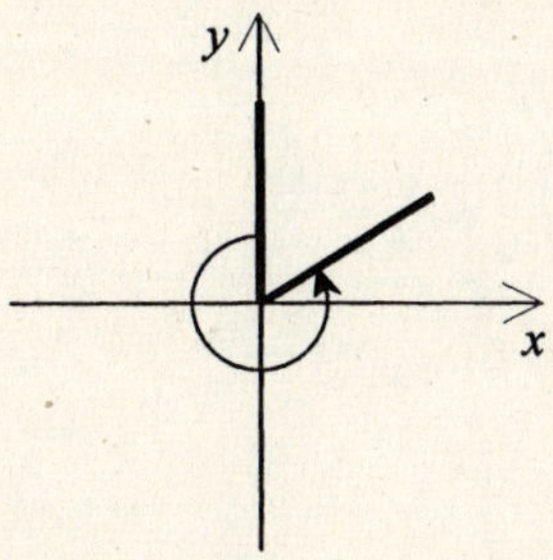

(B)

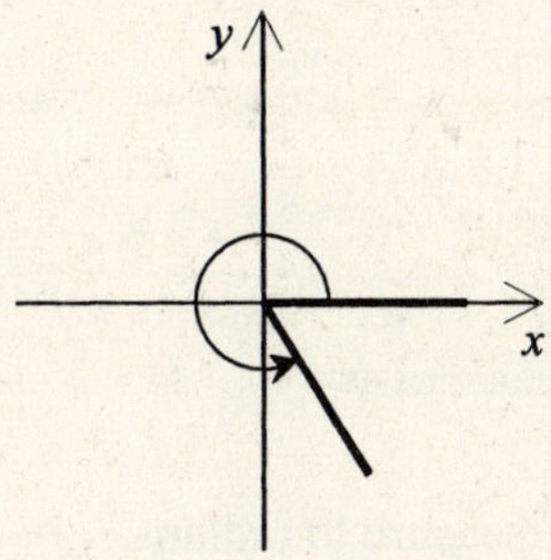

(C)

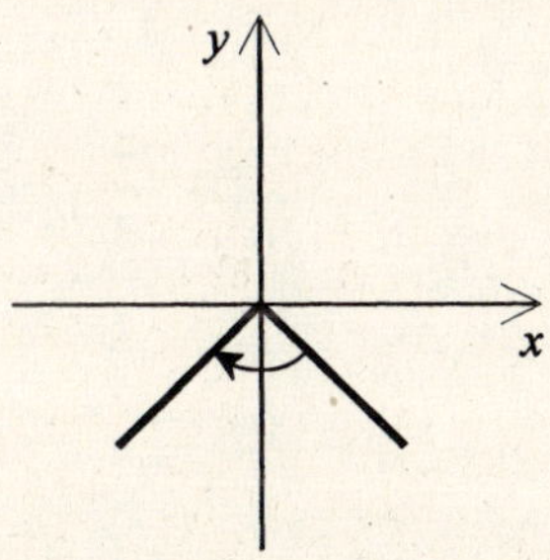

(D)

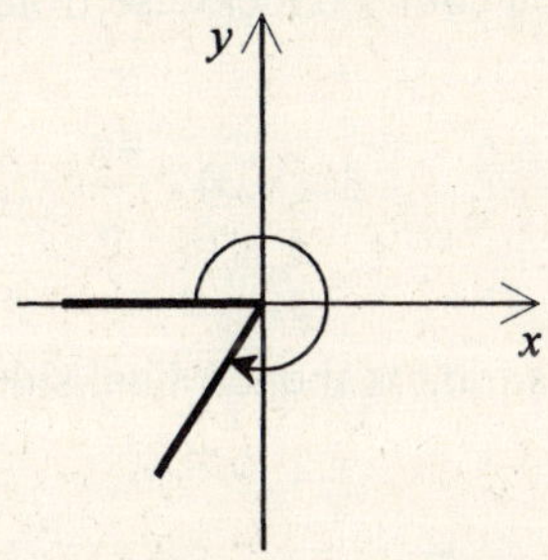

2. Select the correct description of N.

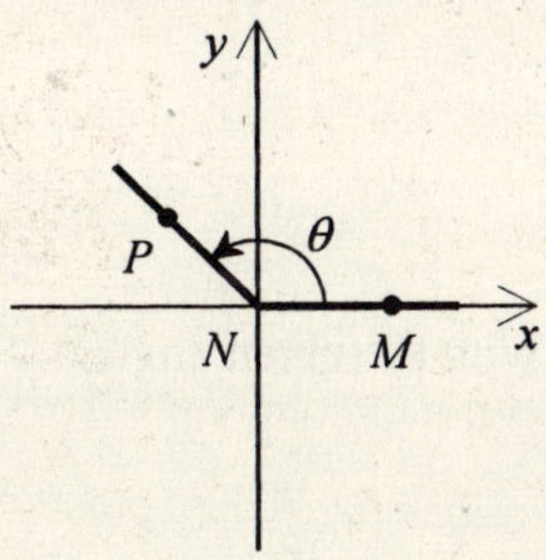

(A) Terminal side of angle θ

(B) Coterminal side of angle θ

(C) Initial side of angle θ

(D) Vertex of angle θ

3. Sketch a negative angle with its terminal side in Quadrant I.

4. Sketch one positive and one negative angle, each of which is coterminal with the angle shown.

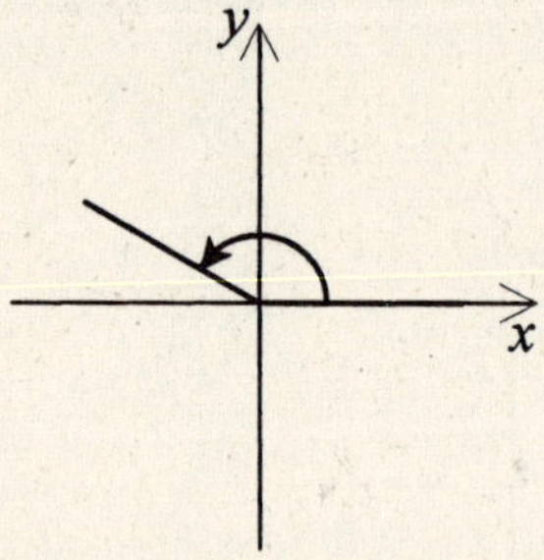

Objective 2: Use radian measure

5. Convert the measure to radians.

$\frac{2}{3}$ revolution counterclockwise from the x-axis

(A) $-\frac{2\pi}{3}$ (B) $\frac{4\pi}{3}$ (C) 240 (D) $\frac{4}{3}$

6. In which quadrant is the terminal side of the angle θ?

$\theta = -\frac{5\pi}{3}$

(A) Quadrant I (B) Quadrant II (C) Quadrant III (D) Quadrant IV

7. Sketch the angle in standard position.

$-\frac{7\pi}{9}$

8. Find one positive angle and one negative angle that are coterminal with the given angle.

$-\frac{\pi}{3}$

Objective 3: Use degree measure

9. Express the angle in radian measure in terms of π. Do not use a calculator.

240°

(A) $\frac{8\pi}{3}$ (B) $\frac{4\pi}{3}$ (C) $\frac{3\pi}{8}$ (D) $\frac{3\pi}{4}$

10. In which quadrant is the terminal side of the angle θ?

$\theta = -55°$

(A) Quadrant I (B) Quadrant II (C) Quadrant III (D) Quadrant IV

11. Express the angle in degree measure.

$\frac{5\pi}{9}$

12. Convert the measure from degrees to radians. Round to three decimal places.

53°

Objective 4: Use angles to model and solve real-life problems

13. The needle of the scale in the bulk food section of a supermarket is 25 cm long. Find the distance the tip of the needle travels when it rotates 50°.

(A) 21.8 cm (B) 125 cm (C) 10.9 cm (D) 3.5 cm

14. A point on the rim of a wheel has a linear speed of 35 cm/s. If the radius of the wheel is 45 cm, what is the angular speed of the wheel in radians per second?

(A) 0.4 rad/s (B) 1.6 rad/s (C) 2.4 rad/s (D) 0.8 rad/s

15. A 50-foot long irrigation sprinkler line rotates around one end as shown. The sprinkler moves through an arc of 250° in 4.55 hours. Find the speed of the moving end of the sprinkler in feet per minute. Round your answer to the nearest tenth.

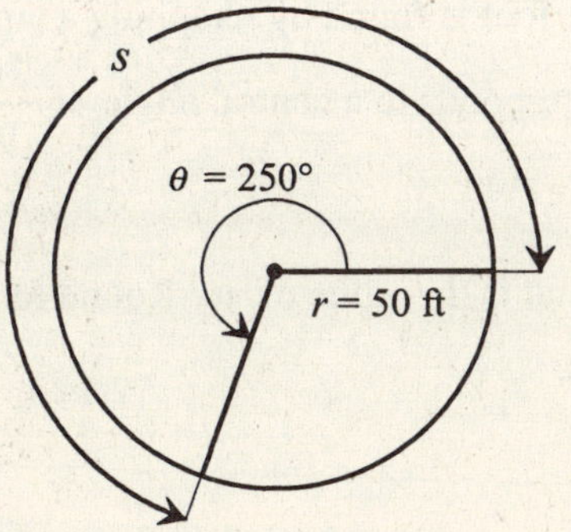

16. One city is 590 miles due north of another city. Assuming that the earth is a sphere of radius 3964 miles, what is the difference in their latitudes?

Section 1.2: Trigonometric Functions: The Unit Circle

Objective 1: Identify a unit circle and its relationship to real numbers

17. Identify the point that lies on the unit circle.

(A) $\left(\frac{17}{13}, -\frac{4}{13}\right)$ (B) $\left(\frac{11}{17}, \frac{6}{17}\right)$ (C) $\left(-\frac{12}{13}, \frac{5}{13}\right)$ (D) $\left(-\frac{4}{9}, -\frac{5}{9}\right)$

18. A number line has its 0 fixed at $(1, 0)$ on a unit circle. What central angle, in radians, is formed by the radius to $(1, 0)$ and the radius to the point that corresponds to 0.97 on the number line when it is wrapped around the circle?

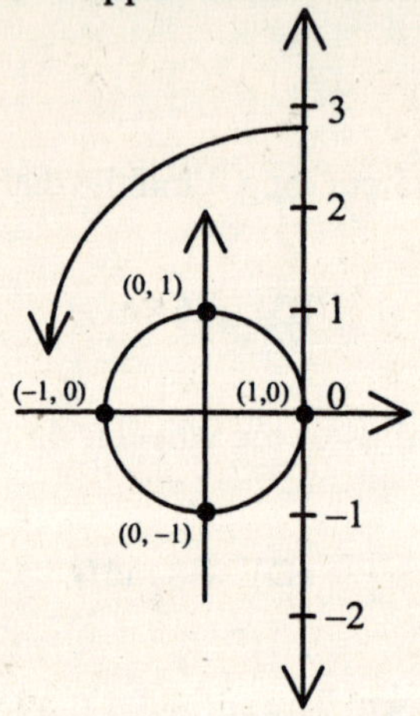

(A) 55.58 (B) $0.97\,\pi$ (C) $55.58\,\pi$ (D) 0.97

19. The length of an arc along a wheel which has a radius of 1 foot is found by wrapping a tape measure around the wheel. What is the arc length that corresponds to a central angle of $\frac{2\pi}{9}$?

20. If the point $(0.19, y)$ is on the unit circle in Quadrant I, what is the value of y? Round to three decimal places.

Objective 2: Evaluate trigonometric functions using the unit circle

21. Use the unit circle and a straightedge to approximate the value of the expression.

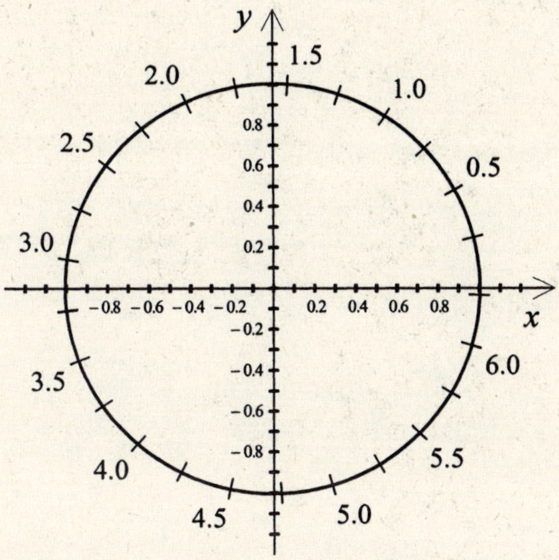

$\sin(-2)$

(A) -0.91 (B) -0.42 (C) 0.91 (D) 0.42

Use the unit circle and symmetry to help you evaluate the function(s).

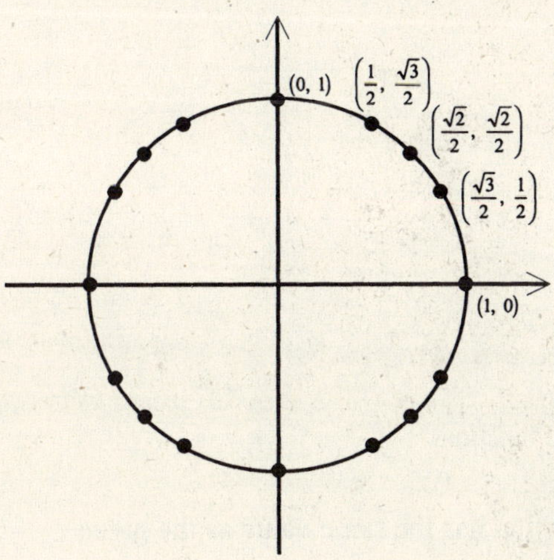

22. $\cot\frac{\pi}{3}$ (A) $\frac{\sqrt{3}}{3}$ (B) $\frac{1}{2}$ (C) $\frac{\sqrt{3}}{2}$ (D) $-\frac{\sqrt{3}}{3}$

Use the unit circle and symmetry to help you evaluate the function(s).

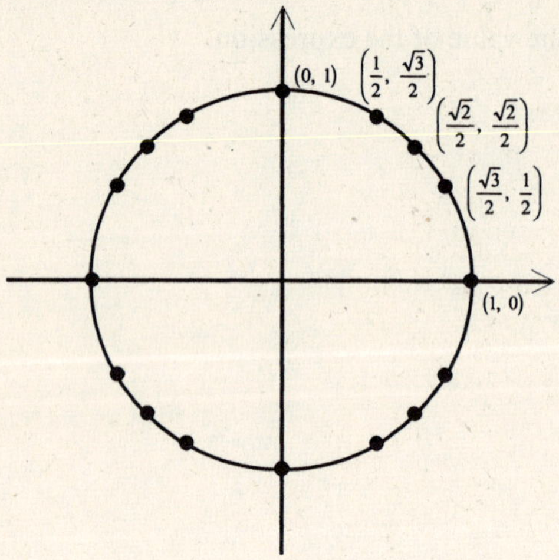

23. All six trigonometric functions of the real number $t = \dfrac{\pi}{4}$

24. Use the coordinates of a point on the unit circle to write the trigonometric ratio.

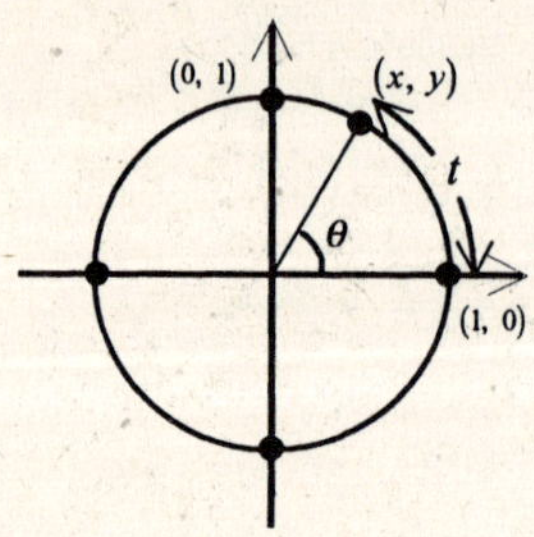

sec t

Objective 3: Use the domain and period to evaluate sine and cosine functions

25. Use the period of the function to select the expression that has the same value as the given expression.

$\sin \dfrac{7\pi}{3}$

(A) $\sin \dfrac{\pi}{3}$ (B) $\sin \dfrac{4\pi}{3}$ (C) $\sin \dfrac{2\pi}{3}$ (D) $\sin \dfrac{\pi}{6}$

26. Use the given trigonometric value to evaluate the indicated expression.
If $\sin t = -\frac{2}{5}$, find $\sin(t + 4\pi)$.

(A) $-\frac{5}{2}$ (B) $-\frac{2}{5}$ (C) $\frac{2}{5}$ (D) $\frac{3}{5}$

27. Use the period of the function to identify an angle in the interval $[0, \pi]$ that generates the same value for the function. Show your reasoning and give the value of the function.
$\cos \frac{33\pi}{4}$

28. Identify the trigonometric function as an even function or an odd function.
$y = \tan\theta$

Objective 4: Use a calculator to evaluate trigonometric functions

Use a calculator to evaluate the expression.

29. $\sin(-1.4)$ (A) -40.9296 (B) -0.0244 (C) -1.0148 (D) -0.9854

30. $\csc 2.5$ (A) 0.5985 (B) -1.0458 (C) 1.6709 (D) 3.3418

31. Use a calculator to evaluate the cotangent, secant, and cosecant of the real number. Round to four decimal places.
$t = 0.5$

32. Use a calculator to evaluate the sine, cosine, and tangent of the real number. Round to four decimal places.
$t = \frac{8\pi}{9}$

Section 1.3: Right Triangle Trigonometry

Objective 1: Evaluate trigonometric functions of acute angles

33. Identify the ratio that defines the trigonometric function of the angle θ.

$\cos\theta$

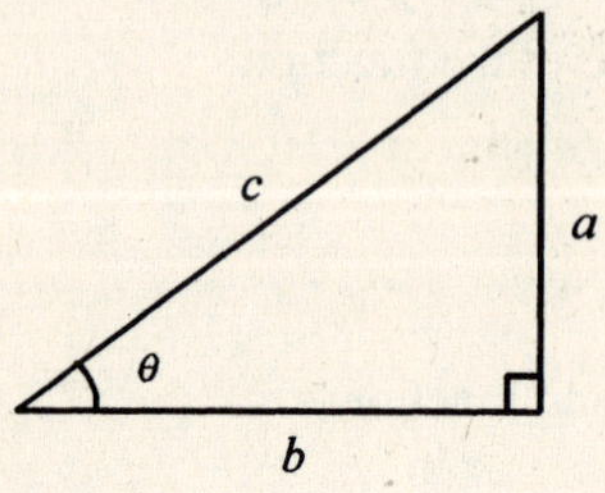

(A) $\frac{a}{b}$ (B) $\frac{c}{a}$ (C) $\frac{b}{c}$ (D) $\frac{c}{b}$

34. Identify the ratio $\frac{b}{c}$ for the indicated angle and find its value.

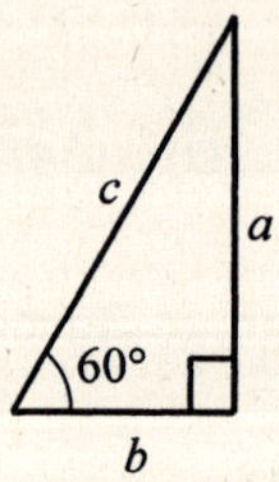

(A) $\sin 60° = \frac{\sqrt{3}}{2}$ (B) $\tan 60° = \sqrt{3}$ (C) $\cos 60° = \frac{1}{2}$ (D) $\sec 60° = \frac{1}{2}$

35. Find the exact value of the six trigonometric functions of the angle θ given in the figure. (Use the Pythagorean Theorem to find the third side of the triangle.)

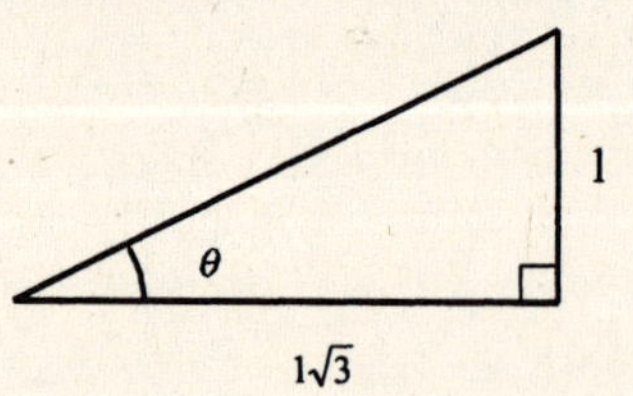

36. Sketch a right triangle corresponding to the trigonometric function of the acute angle θ. Use the Pythagorean Theorem to determine the third side and then find the indicated trigonometric function of θ.

If $\tan\theta = \frac{\sqrt{3}}{3}$, find $\sin\theta$.

Objective 2: Use the fundamental trigonometric identities

37. Let θ be an acute angle. Use the given function value and trigonometric identities to find the indicated trigonometric function.

If $\csc\theta = \frac{41}{9}$, find $\sin\theta$.

(A) $\frac{9}{41}$ (B) $\frac{40}{9}$ (C) $\frac{41}{40}$ (D) $\frac{40}{41}$

38. Use the fundamental trigonometric identities to determine the simplified form of the expression.

$$\frac{\tan\beta}{\sin\beta}$$

(A) $\cos\beta$ (B) $\csc\beta$ (C) $\cot\beta$ (D) $\sec\beta$

39. Use the fundamental trigonometric identities to determine a simplified form of the expression.

$$(1-\sec\beta)(1+\sec\beta)$$

40. Let θ be an acute angle. Use the given function value and trigonometric identities to find the indicated trigonometric function.

If $\cos\theta = \frac{1}{5}$, find $\sin\theta$.

Objective 3: Use a calculator to evaluate trigonometric functions

Use a calculator to evaluate the function. (Be sure the calculator is in the correct angle mode.)

41. $\sec 37^\circ 50' 38''$ (A) 1.2663 (B) 0.7897 (C) 1.3896 (D) 1.0106

42. $\tan 56.22^\circ$ (A) 0.6689 (B) -0.6062 (C) -0.3410 (D) 1.4949

Use a calculator to evaluate the function. Round your answer to four decimal places. (Be sure the calculator is in the correct angle mode.)

43. sec 63.79°

44. cot 29° 12′

Objective 4: Use trigonometric functions to model and solve real-life problems

45. The cable supporting a ski lift rises 3 feet for each 8 feet of horizontal length. The top of the cable is fastened 1087.5 feet above the cable's lowest point. Find the lengths b and c, and find the measure of angle θ.

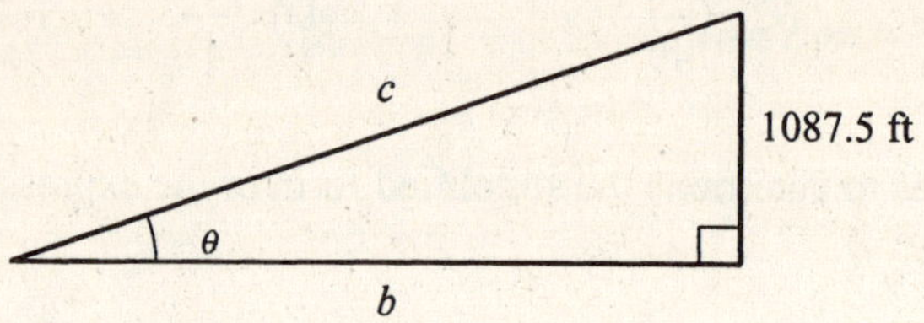

(A) $b = 1161$ ft, $c = 408$ ft, $\theta = 0.4°$
(B) $b = 3097$ ft, $c = 2900$ ft, $\theta = 22.0°$
(C) $b = 408$ ft, $c = 1161$ ft, $\theta = 69.4°$
(D) $b = 2900$ ft, $c = 3097$ ft, $\theta = 20.6°$

46. An 18-foot ladder makes an angle of 55° with the ground as it leans against a wall. How far up the wall does the ladder reach?

(A) 14.74 ft (B) 25.71 ft (C) 21.97 ft (D) 10.32 ft

47. A photographer points a camera at a window in a nearby building forming an angle of 40° with the camera platform. If the camera is 59 meters from the building, how high above the platform is the window? Round to two decimal places.

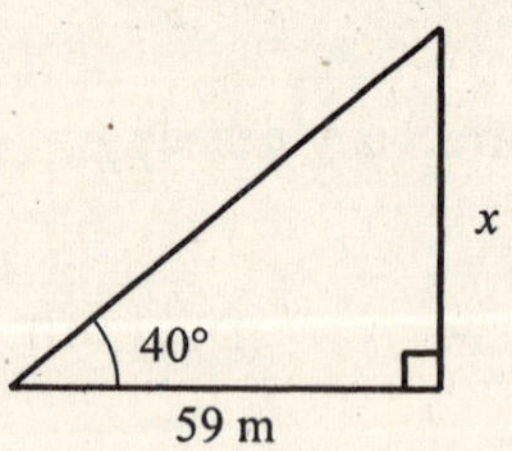

48. The tailgate of a truck is $2\frac{3}{4}$ feet above the ground. The incline of a ramp used for loading the truck is 20°. Find the length of the ramp. Round your answer to one decimal place.

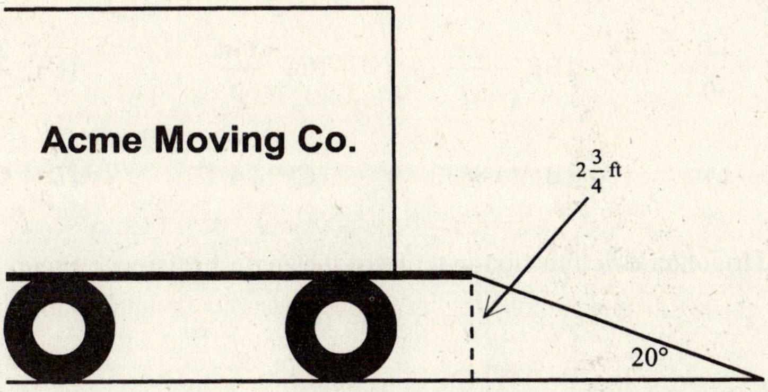

Section 1.4: Trigonometric Functions of Any Angle

Objective 1: Evaluate trigonometric functions of any angle

49. Given $\tan\theta = -\frac{5}{8}$ and $\sin\theta > 0$, find $\cos\theta$.

(A) $\cos\theta = -\frac{8\sqrt{89}}{89}$ (B) $\cos\theta = \frac{8\sqrt{89}}{89}$ (C) $\cos\theta = -\frac{5\sqrt{89}}{89}$ (D) $\cos\theta = \frac{5\sqrt{89}}{89}$

50. Identify the quadrant in which θ lies.
$\sin > 0$ and $\tan < 0$

(A) Quadrant I (B) Quadrant II (C) Quadrant III (D) Quadrant IV

51. Evaluate the sine, cosine, and tangent functions of the quadrant angle.
$\theta = \frac{3\pi}{2}$

52. The point given is on the terminal side of an angle in standard position. Find the cotangent, the secant, and the cosecant of the angle.
$(-7, 24)$

Objective 2: Use reference angles to evaluate trigonometric functions

Find the reference angle θ'.

53. $\theta = -\frac{7\pi}{9}$ (A) $-\frac{11\pi}{9}$ (B) $-\frac{2\pi}{9}$ (C) $\frac{11\pi}{9}$ (D) $\frac{2\pi}{9}$

54. $\theta = -1.4$ (A) -1.4 (B) 4.8832 (C) 1.4 (D) 0.1708

55. Rewrite the indicated trigonometric function in terms of the angle's reference angle. Use the same function.
$\sin(-418°)$

56. Angle θ is shown along with the value of its reference angle. Rewrite the indicated function in terms of the reference angle θ' and give the value of the function. Round your answer to four decimal places.
$\tan\theta$

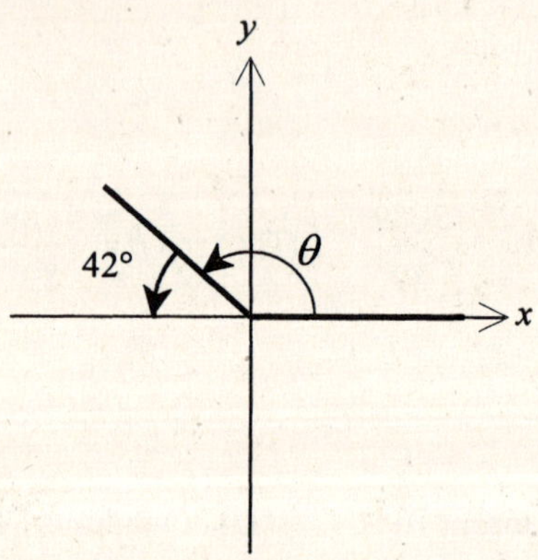

Objective 3: Evaluate trigonometric functions of real numbers

57. Find the exact value of the function. (A) 1 (B) $-\frac{\sqrt{3}}{3}$ (C) $\sqrt{3}$ (D) 2
$\csc 150°$

58. Use a calculator to approximate two values of $\theta\ (0 \le \theta < 2\pi)$ that satisfy the equation.
$\sin\theta = -0.9825$

(A) 5.685, 5.310 (B) 4.900, 4.525 (C) 4.900, 5.310 (D) 5.685, 4.525

59. Find the exact value of the function.

$\tan \frac{19\pi}{6}$

60. Find the indicated trigonometric value in the specified quadrant.

θ is in Quadrant IV and $\cos\theta = \frac{1}{2}$. Find $\sin\theta$.

Section 1.5: Graphs of Sine and Cosine Functions

Objective 1: Sketch the graphs of basic sine and cosine functions

61. Give the number of full cycles of the function that are found in the interval.
$y = -\cos x$ on the interval $[-\pi,\ 15\pi]$

(A) 7 (B) 9 (C) 8 (D) 16

62. Identify the function shown in the graph.

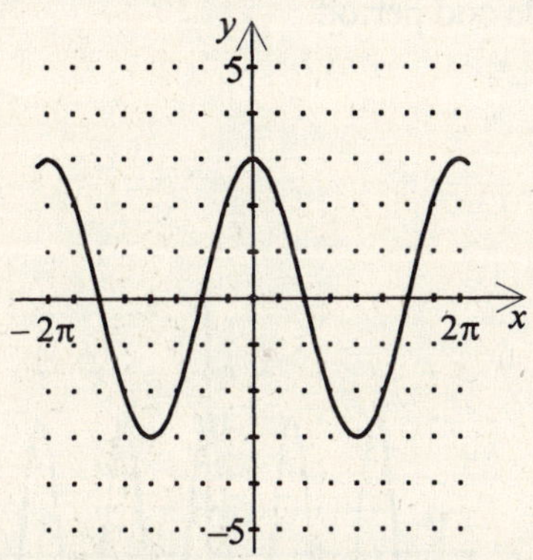

(A) $y = \cos 3x$ (B) $y = \sin 3x$ (C) $y = 3\sin x$ (D) $y = 3\cos x$

63. List the key points on the graph of the function.
$y = 2.5\cos x$ on the interval $[0, 2\pi]$

64. Sketch the graph of the function.
$y = 1.5\sin x$ on the interval $[-2\pi, 2\pi]$

Objective 2: Use amplitude and period to help sketch the graphs of sine and cosine functions

65. Find the amplitude and the period of the function.

$y = 0.5\cos\dfrac{x}{3}$

(A) Amplitude = 0.5
Period = 6π

(B) Amplitude = – 0.5
Period = 6π

(C) Amplitude = – 0.5
Period = $\dfrac{1}{3}$

(D) Amplitude = 0.5
Period = $\dfrac{1}{3}$

66. Identify the function that has the given amplitude and period.

Amplitude = 1.5 , period = $\dfrac{3\pi}{4}$

(A) $y = 1.5\sin\dfrac{8x}{3}$ (B) $y = 0.75\sin\dfrac{8x}{3}$ (C) $y = 1.5\sin\dfrac{4\pi x}{3}$ (D) $y = 0.75\sin\dfrac{3\pi x}{4}$

67. Sketch a graph of the function that has the given amplitude and period.

Cosine function, amplitude = 2 , period = $\dfrac{\pi}{2}$

68. Find the amplitude and the period of the graphed function.

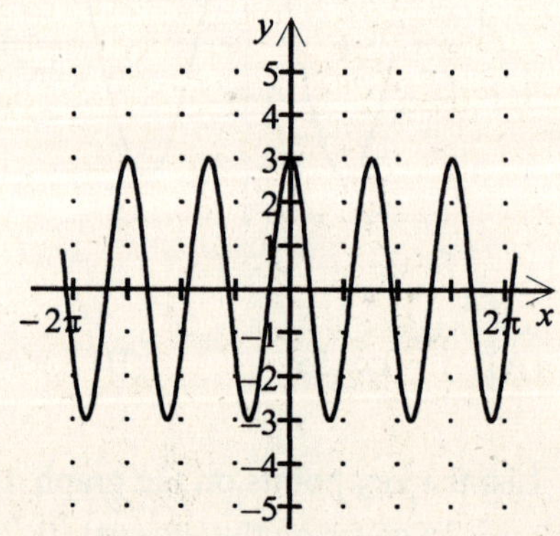

Objective 3: Sketch translations of the graphs of sine and cosine functions

69. Identify the graph of the function.
$y = 1.5\cos 2x - 2$

(A)

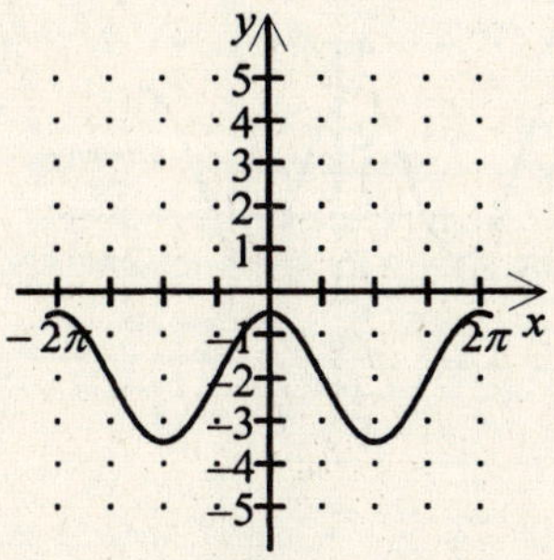

(B)

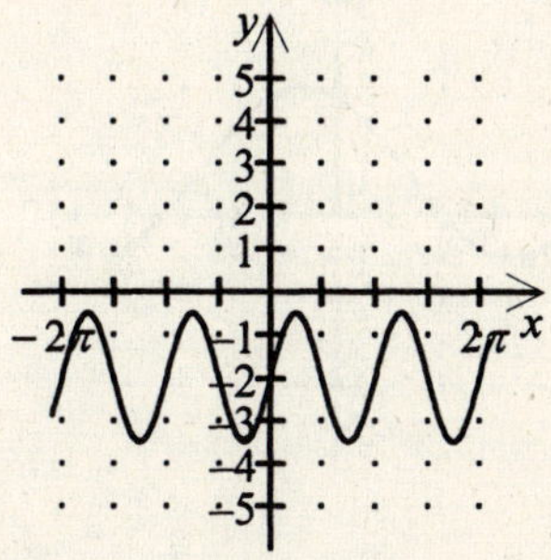

(C)

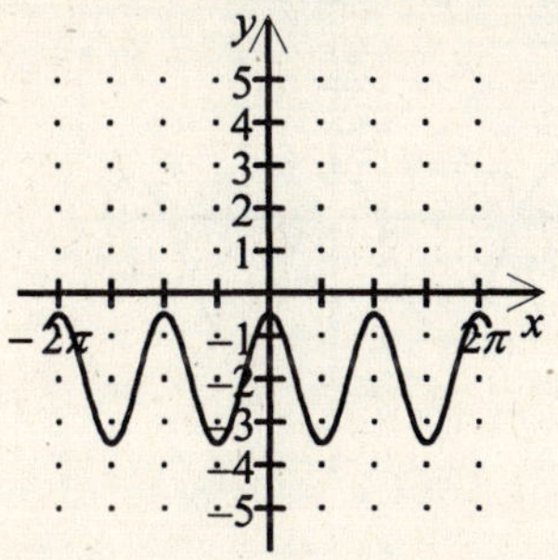

(D)

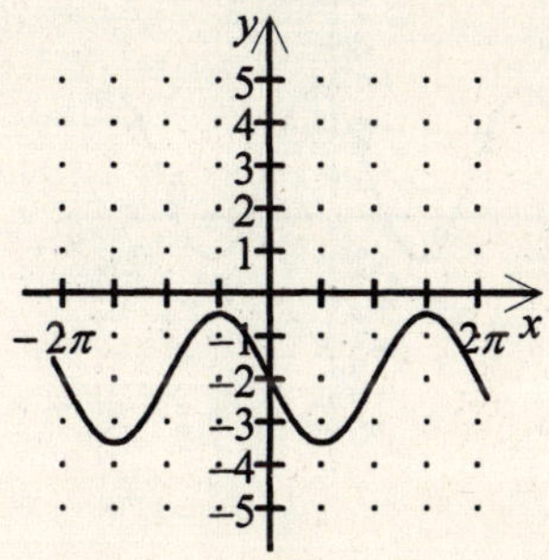

70. Identify the graph of the cosine function that has the required phase shift and vertical translation.

phase shift $= -\frac{\pi}{6}$; vertical shift $= 1$

(A)

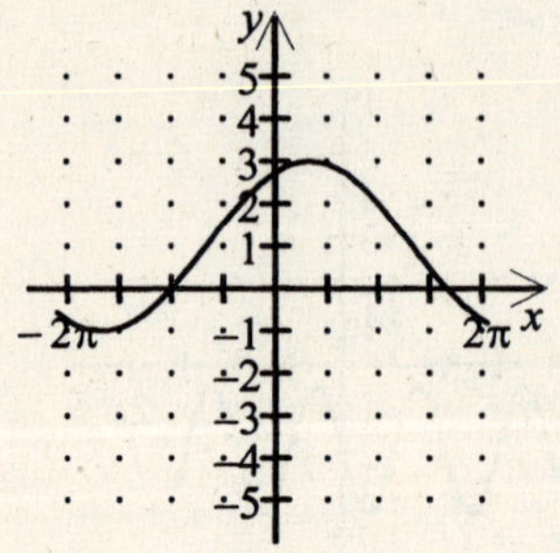

(B)

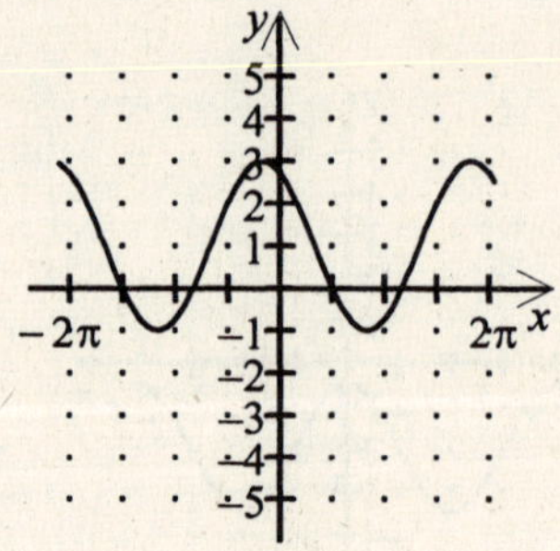

(C)

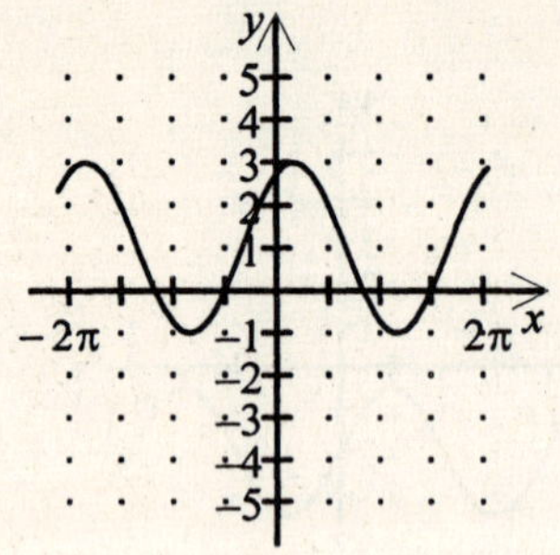

(D)

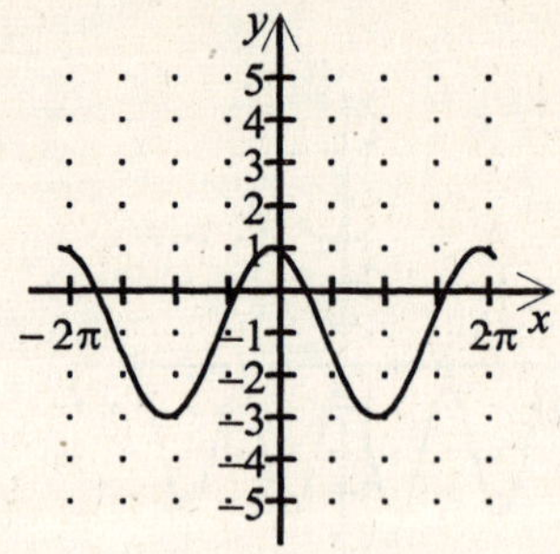

71. List the start and end points of any full cycle on the graph of the function. $y = -2\sin(6x + 5\pi)$

72. Sketch the graph of the function.

$y = -\sin\left(x + \frac{\pi}{4}\right) + 1$, on the interval $[-2\pi, 2\pi]$

Objective 4: Use sine and cosine functions to model real-life data

73. The data below represent the average monthly cost of natural gas in an Oregon home.

Month	Aug	Sep	Oct	Nov	Dec	Jan
Cost ($)	20.60	27.06	45.23	68.35	93.47	110.14

Month	Feb	Mar	Apr	May	Jun	Jul
Cost ($)	117.10	110.64	92.97	68.85	44.73	27.06

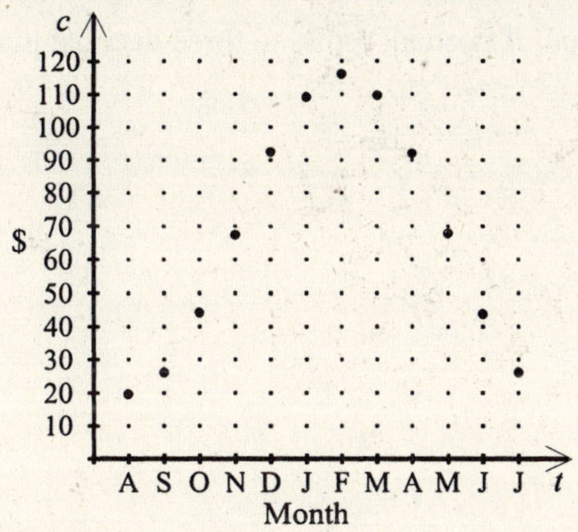

Which sine function best describes the data?

(A) $c(t) = 48.3 \sin\left(\frac{\pi t}{6} - \frac{\pi}{12}\right) + 68.8$

(B) $c(t) = 48.3 \sin\left(\frac{\pi t}{6} - \frac{2\pi}{3}\right) + 68.8$

(C) $c(t) = 48.3 \sin\left(\frac{\pi t}{8} - 12\right) + 20.6$

(D) $c(t) = 48.3 \sin\left(\frac{\pi t}{4} - \frac{2\pi}{3}\right) + 20.6$

74. A weight attached to the end of a spring is pulled down 8 centimeters below its equilibrium point and released. It takes 5 seconds for it to complete one cycle of moving from 8 centimeters below the equilibrium point to 8 centimeters above the equilibrium point and then returning to its low point. Identify the sinusoidal function that best represents the position of the moving weight and the approximate position of the weight at 8 seconds.

(A) $h(t) = 5\cos\frac{\pi t}{4}$, 1 cm

(B) $h(t) = 8\sin\frac{2\pi t}{5}$, 6.5 cm

(C) $h(t) = -5\sin\frac{\pi t}{4}$, 5 cm

(D) $h(t) = -8\cos\frac{2\pi t}{5}$, 6.5 cm

75. Krysta is checking her bicycle on a stand. The bicycle's wheel has a radius of 31 centimeters and its center is 57 centimeters above the ground. Krysta is rotating the wheel at a constant angular speed of $\frac{5\pi t}{3}$ radians per second.

(a) Describe the function $h(t)$ using the sine function where h is the height in feet and t is the time in minutes.

(b) Find $h(t)$ when $t = 13$ seconds. Round your answer to the nearest tenth of a centimeter.

76. The high and low tides of a coastal city on May 3, 2000 are shown in the table. Find an equation for $h(t)$, the height of the water as a function of t, the number of hours after midnight. Assume that the water height can be modeled by a cosine function. If needed, round to three decimal places.

Tide	Time	Height
High	3:00 A.M.	108 cm
Low	9:24 A.M.	52 cm

Section 1.6: Graphs of Other Trigonometric Functions

Objective 1: Sketch the graphs of tangent functions

77. Identify the graph of the function.

$$y = -\tan\frac{5x}{3}$$

(A)

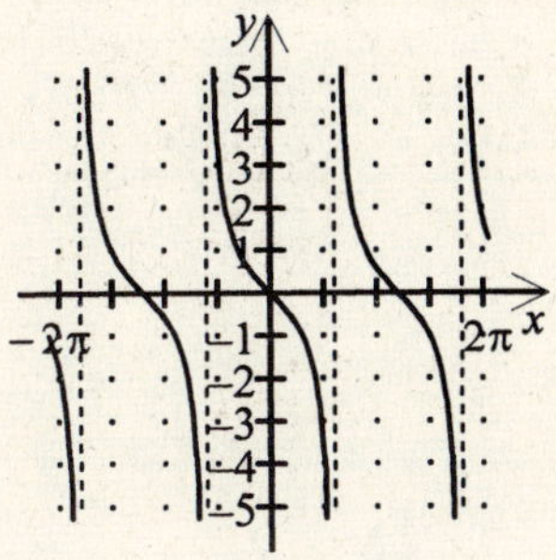

(B)

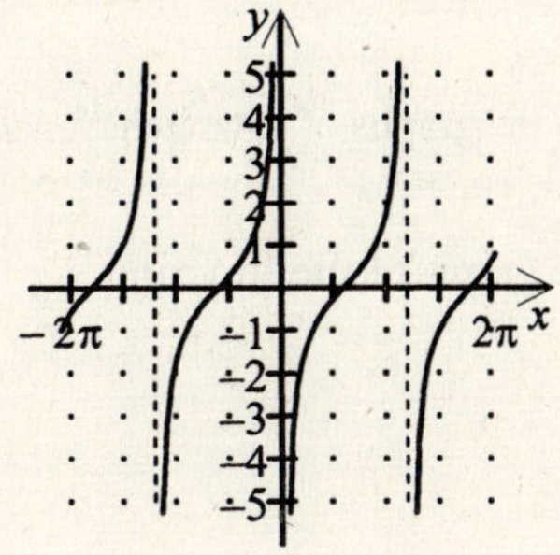

(C)

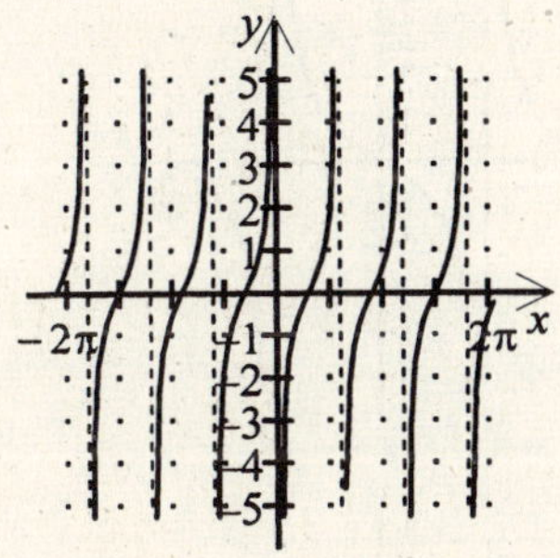

(D)

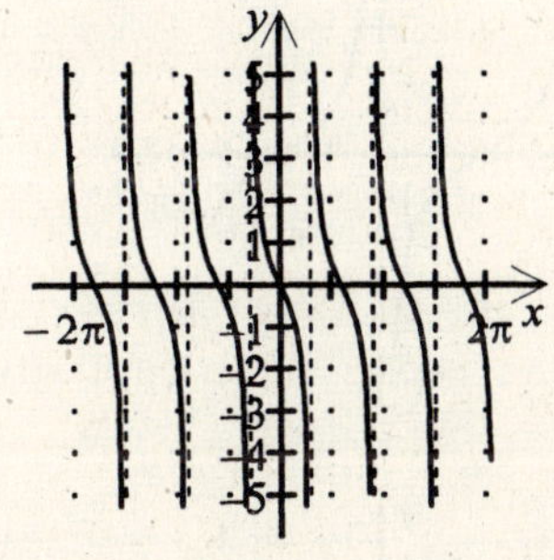

78. Identify the function that is graphed.

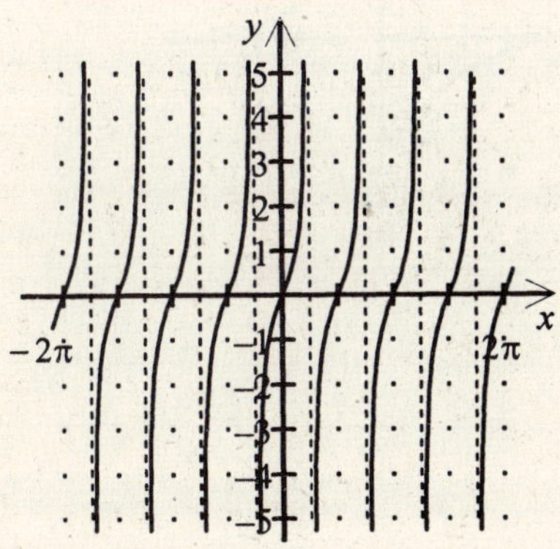

(A) $y = -\tan\frac{x}{2}$ (B) $y = \cot\frac{x}{2}$ (C) $y = \tan 2x$ (D) $y = -\cot 2x$

79. Sketch the graph of the function and write equations for 2 consecutive asymptotes.

$$y = -\tan\left(\frac{9x}{4} + \frac{9\pi}{16}\right)$$

80. Sketch the graph of the function. Include any asymptotes.

$$y = -0.5\tan\frac{3\pi x}{10}$$

Objective 2: Sketch the graphs of cotangent functions

81. Identify the graph of the function.

$$y = \cot\frac{x}{2}$$

(A)

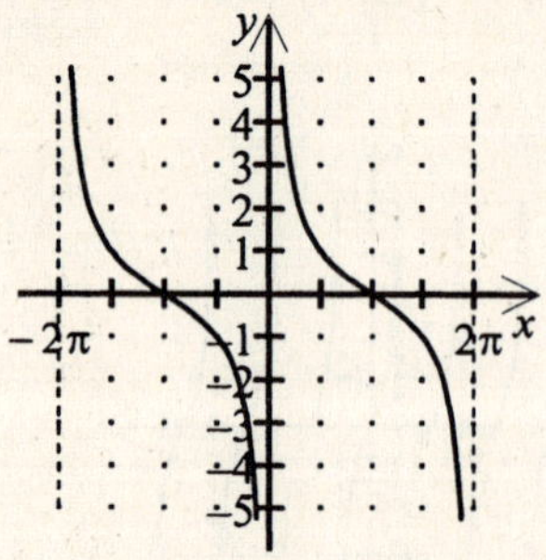

(B)

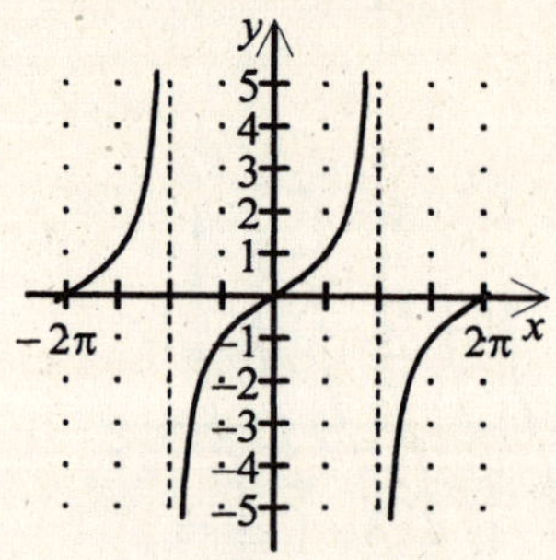

(C)

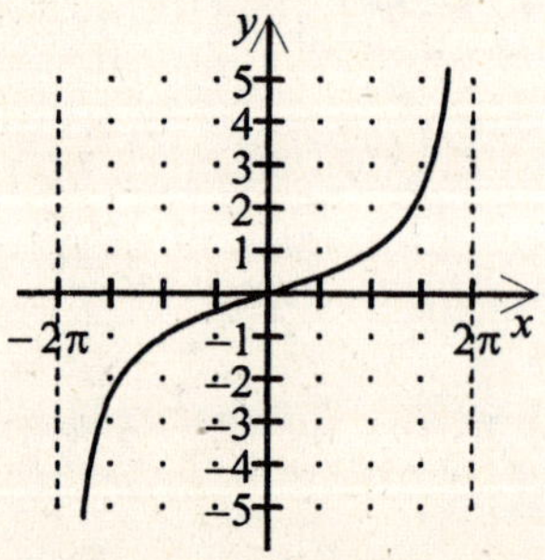

(D)

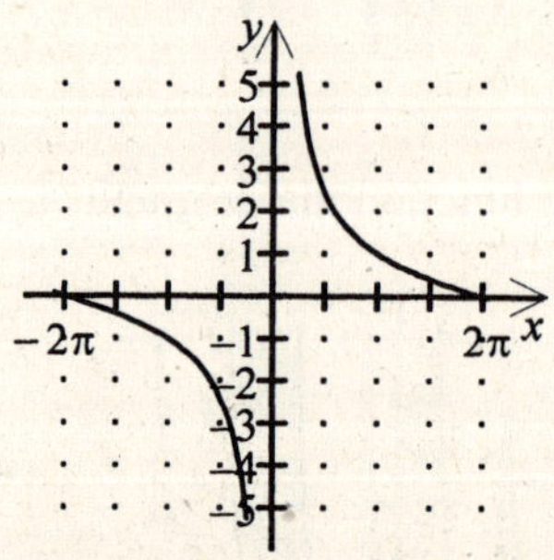

82. Identify the function that is graphed.

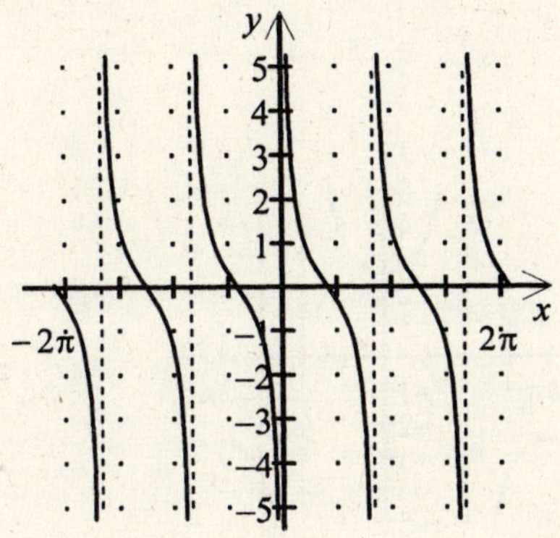

(A) $y = \cot \dfrac{6x}{5}$ (B) $y = \tan \dfrac{5x}{6}$ (C) $y = -\tan \dfrac{6x}{5}$ (D) $y = -\cot \dfrac{5x}{6}$

83. Sketch the graph of the function and write equations for 2 consecutive asymptotes.

$$y = \cot\left(\frac{9x}{4} + \frac{9\pi}{8}\right)$$

84. Sketch the graph of the function. Include any asymptotes.

$$y = 2.5 \cot \frac{3\pi x}{10}$$

Objective 3: Sketch the graphs of secant and cosecant functions

85. Identify the equation that matches the description of the function.

A secant function has a phase shift of $-\dfrac{5\pi}{4}$ and a period of $\dfrac{10\pi}{3}$.

(A) $y = \sec\left(\dfrac{3x}{5} - \dfrac{5\pi}{4}\right)$ (B) $y = \sec\left(\dfrac{10x}{3} + \dfrac{3\pi}{4}\right)$

(C) $y = \sec\left(\dfrac{3x}{5} - \dfrac{3\pi}{4}\right)$ (D) $y = \sec\left(\dfrac{3x}{5} + \dfrac{3\pi}{4}\right)$

86. Identify the graph of the function.

$$y = 2.5\csc\frac{x}{2}$$

(A)

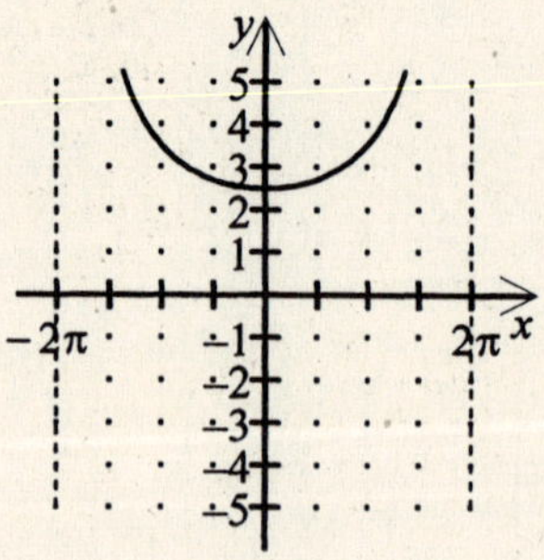

(B)

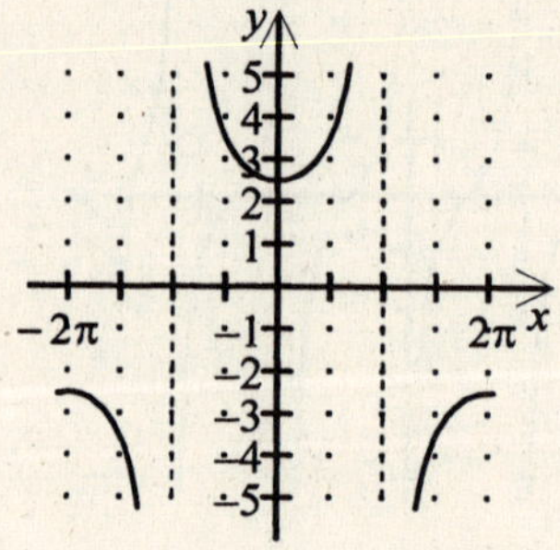

(C)

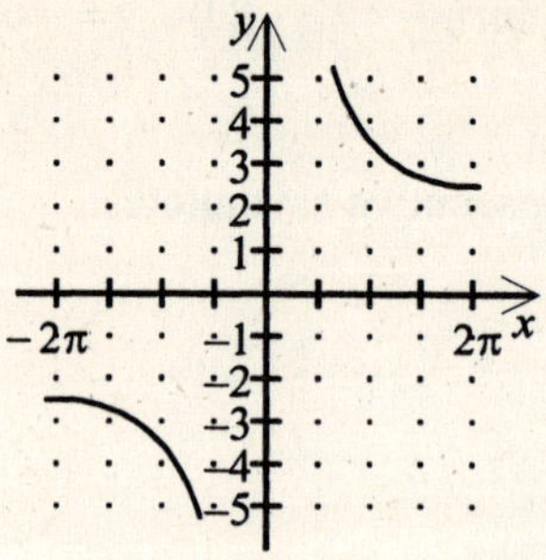

(D)

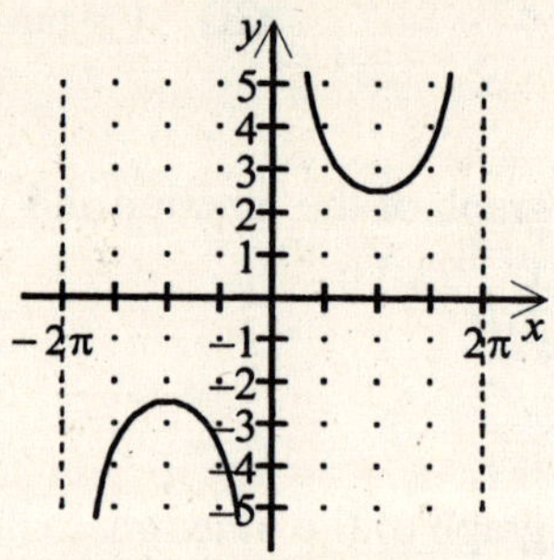

87. Sketch the graph of the function and write equations for 2 consecutive asymptotes.

$$y = 1.5\sec\left(\frac{5x}{3} + \frac{5\pi}{12}\right)$$

88. Give the period and the phase shift of the function.

$$y = -3.5\sec\left(\frac{x}{3} + \frac{\pi}{3}\right)$$

Objective 4: Sketch the graphs of damped trigonometric functions

89. Identify the graph of the function.
$y = |2x| \cos 4x$

(A)

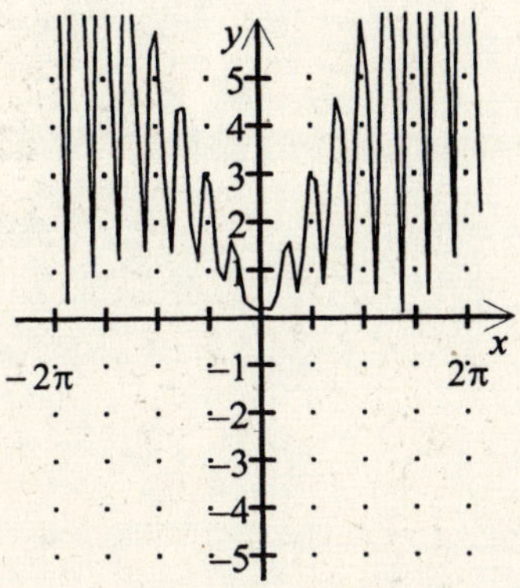

(B)

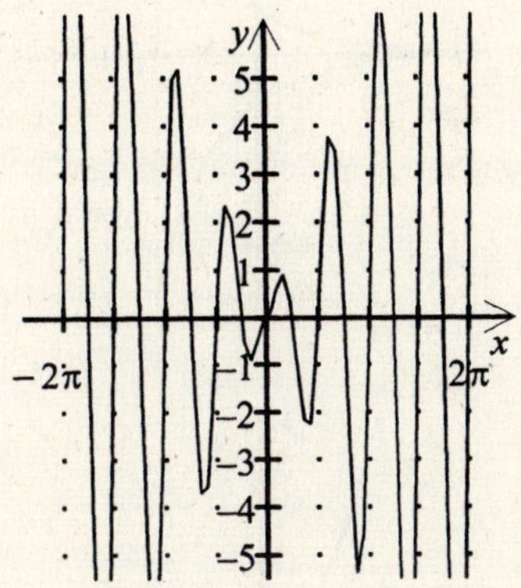

(C)

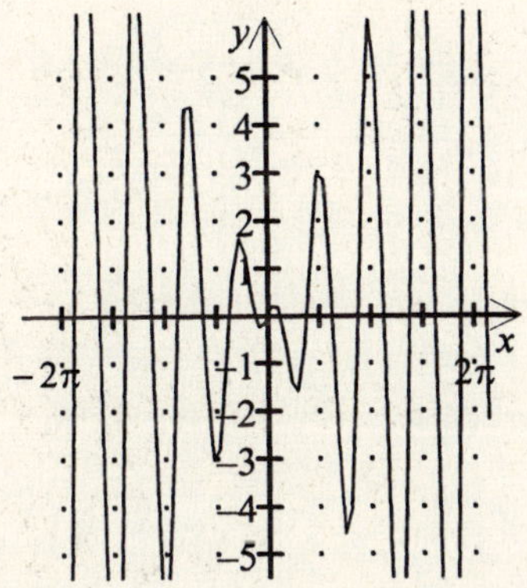

(D)

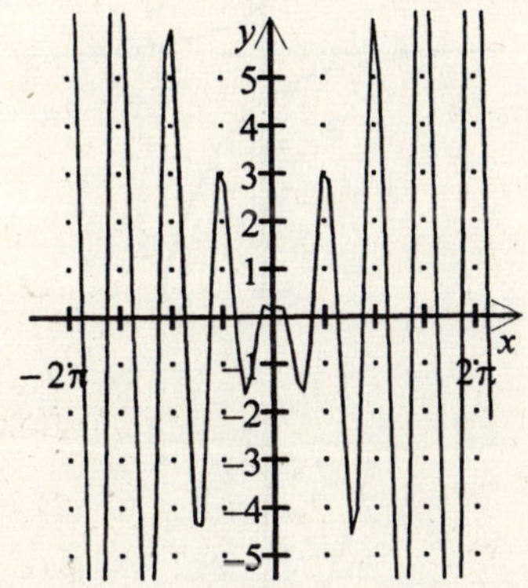

90. Identify the graph that shows the boundaries outlining the maximum and minimum values of the function.
$y = e^{-x/2} \cos x$

(A)

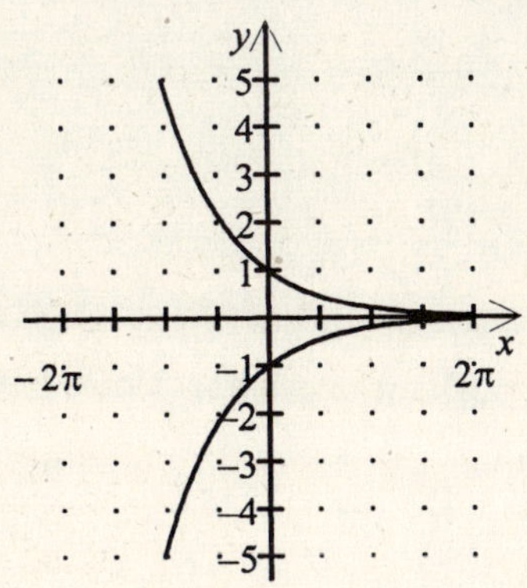

(B)

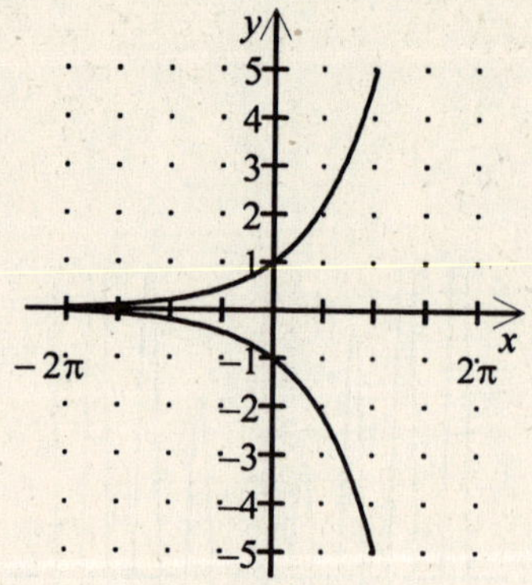

(C)

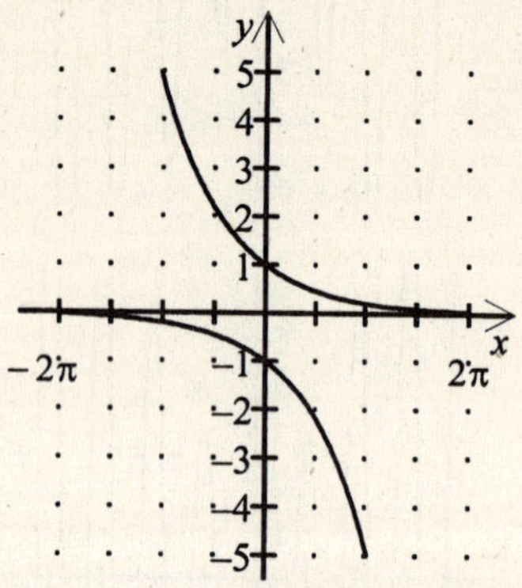

(D)

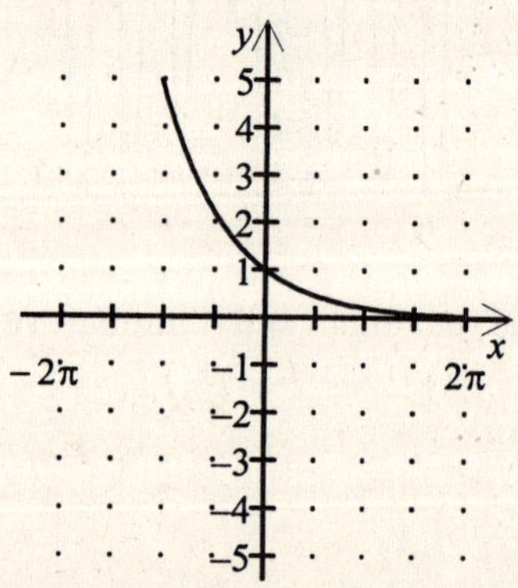

(90.)

91. What is the damping factor in the equation?

$$f(x) = \left(\frac{e}{2}\right)^{-x/4} \sin 4x$$

92. Use a graphing utility to sketch the graph of the function.

$$y = \frac{\cos 2x}{|2x|}$$

Section 1.7: Inverse Trigonometric Functions

Objective 1: Evaluate the inverse sine function

93. If possible, evaluate the expression without the aid of a calculator.

$$\sin^{-1}\left(-\frac{\sqrt{3}}{2}\right)$$

(A) $-\frac{\pi}{6}$ (B) $-\frac{\pi}{3}$ (C) $-\frac{\pi}{4}$ (D) Not possible

94. Use a calculator to approximate the expression.
arcsin 0.26

(A) 220.37 (B) 0.05 (C) 3.89 (D) 0.26

95. Determine the exact value of the missing coordinate of the point marked on the graph of the function.

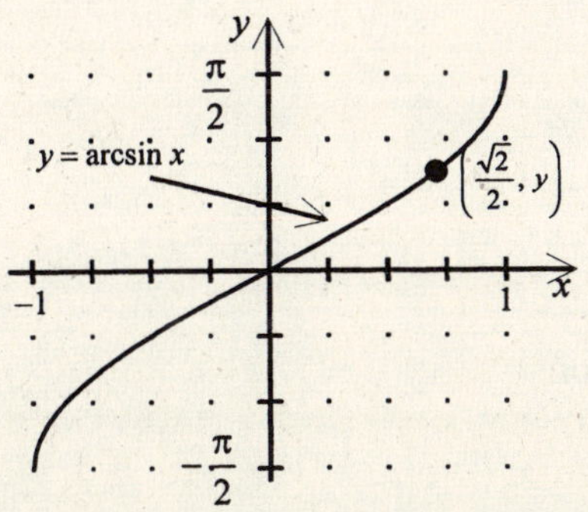

96. Give the exact value in radians.
$\theta = \arcsin 0$

Objective 2: Evaluate the other inverse trigonometric functions

Use a calculator to approximate the expression.

97. $\arccos(-0.15)$ (A) 1.00 (B) 1.72 (C) 1.57 (D) 1.01

98. $\arctan 0.74$ (A) 1.0952 (B) 0.6371 (C) 0.0129 (D) 77.4224

99. Give the exact value in radians.

$\theta = \arccos \frac{\sqrt{3}}{2}$

100. Determine the exact value of the missing coordinate of the point indicated on the graph of the function.

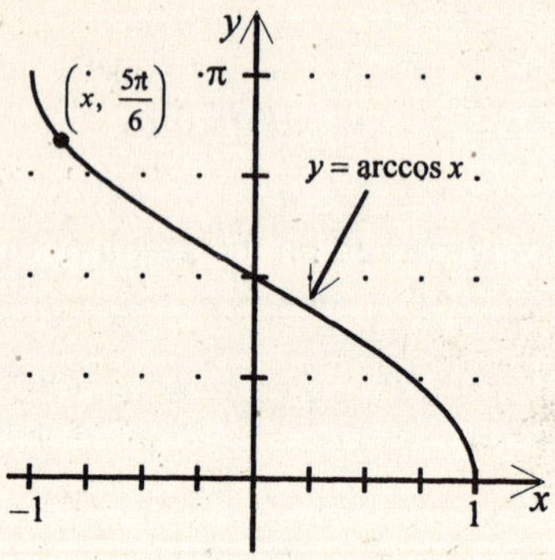

Objective 3: Evaluate the compositions of trigonometric functions

Use the properties of inverse functions to evaluate the expression.

101. $\csc\left(\arcsin \frac{5}{13}\right)$ (A) $\frac{13}{5}$ (B) $\frac{12}{13}$ (C) $\frac{5}{12}$ (D) $\frac{5}{13}$

102. $\sec\left(\arccos \frac{x}{2}\right)$ (A) $\frac{2}{x}$ (B) $\frac{x}{2}$ (C) $\frac{2\sqrt{4-x^2}}{4-x^2}$ (D) $\frac{\sqrt{4-x^2}}{x}$

103. $\arccos\left(\cos\left(-\frac{13\pi}{8}\right)\right)$

Use the properties of inverse functions to evaluate the expression.

104. $\sin\left(\arctan\dfrac{\sqrt{3}}{3}\right)$

Section 1.8: Applications and Models

Objective 1: Solve real-life problems involving right triangles

105. At a distance of 24 feet from the base of a flag pole, the angle of elevation to the top of a flag that is 4.1 feet tall is 52.8°. The angle of elevation to the bottom of the flag is 48.9°. The pole extends 1 foot above the flag. Find the height of the pole.

(A) 32.6 ft (B) 28.8 ft (C) 27.8 ft (D) 31.6 ft

106. An airplane is flying east at a constant altitude of 23,500 meters. When first seen to the east of an observer, the angle of elevation to the airplane is 68.6°. After 32 seconds, the angle of elevation is 59.7°. Find the speed of the airplane.

(A) 270 m/s (B) 142 m/s (C) 136 m/s (D) 286 m/s

107. A flat roof rises at a 24° angle from the front wall of a storage shed to the back wall. The front wall is 9.5 feet tall and the back wall is 13.5 feet tall. Find the length of the roof line and the depth of the shed from front to back. Round your answers to the nearest tenth of a foot.

108. The pilot of a small private plane can look forward and see the control tower for a small airstrip. Beyond that is a large factory that is 3.3 miles from the airstrip. The angles of depression are 7° and 3.6°, respectively. Find the airplane's altitude, to the nearest ten feet.

Objective 2: Solve real-life problems involving directional bearings

109. A ship leaves port at 12 miles per hour, with a heading of N 62° E. There is a warning buoy located 5 miles directly north of the port. What is the bearing of the warning buoy as seen from the ship after 7.5 hours?

(A) N 64.9° W (B) S 25.1° E (C) S 64.9° W (D) N 25.1° E

110. A hiker travels at 2.8 miles per hour at a heading of N 24° E from a ranger station. After 7.5 hours, how far north and how far east is the hiker from the ranger station?

(A) 8.9 miles north and 19.0 miles east (B) 19.0 miles north and 8.9 miles east

(C) 8.5 miles north and 19.2 miles east (D) 19.2 miles north and 8.5 miles east

111. A private seaplane travels in a straight line from a local airstrip to a remote lake that is 370 miles east and 443 miles north of the airstrip. Find the lake's bearing and distance from the airstrip. Round your answers to the nearest tenth.

112. What direction bearing is shown?

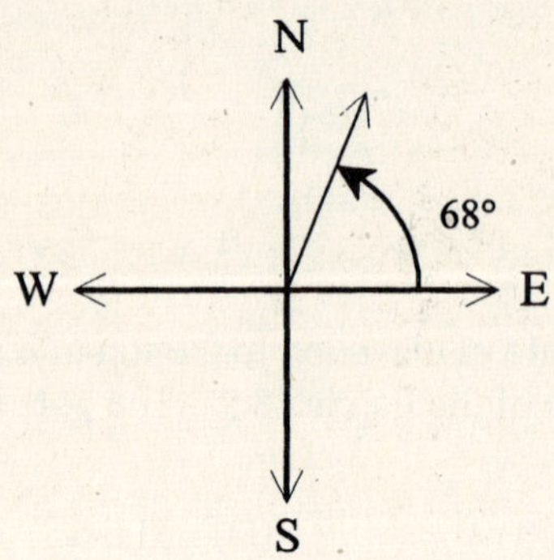

Objective 3: Solve real-life problems involving harmonic motion

113. A rowboat is observed from a dock as it bobs up and down in simple harmonic motion because of wave action. The boat moves from a high point of 3.9 feet below the dock to a low point of 4.5 feet below the dock and back to its high point 9 times every minute. Let t be time, in minutes, and h be the distance below the dock, in feet. Identify the equation that describes the boat's motion.

(A) $h(t) = -4.2 + 0.3\cos\frac{2\pi}{9}t$

(B) $h(t) = 0.6\sin 18\pi t + 4.2$

(C) $h(t) = -0.6\sin\frac{2\pi}{9}t$

(D) $h(t) = 4.2 - 0.3\cos 18\pi t$

114. A mass attached to a spring vibrates up and down in simple harmonic motion according to the equation

$$h(t) = -6.5\sin\frac{3\pi}{4}t$$

where h is in centimeters and t is in seconds. Identify the amplitude and the period of the vibrations.

(A) Amplitude = 6.5 cm
Period = $\frac{8\pi}{3}$ sec

(B) Amplitude = 6.5 cm
Period = $\frac{8}{3}$ sec

(C) Amplitude = 6.5 cm
Period = $\frac{3\pi}{4}$ sec

(D) Amplitude = – 6.5 cm
Period = $\frac{3\pi}{4}$ sec

115. If you watch from ground level, a child riding on a merry-go-round will seem to be undergoing simple harmonic motion from side to side. Assume the merry-go-round is 12.6 feet across and the child completes 6 rotations in 90 seconds. Write a sine function that describes *d*, the child's apparent distance from the center of the merry-go-round, as a function of time *t*.

116. If a pendulum bob is released from its farthest right position and it swings in small arcs, then the distance *x* from the equilibrium position can be described by a cosine function. Assume a pendulum bob swings in an arc 10 centimeters wide with a period of 1.9 seconds. Let *x* be positive if the bob is to the right of the equilibrium position and let *t* be the time after the bob's release. What is the equation for the bob's horizontal location and where is the bob after 2.5 seconds? Round as needed to two decimal places.

Answer Key for Chapter 1 Trigonometry

Section 1.1: Radian and Degree Measure

Objective 1: Describe angles

[1] (B)

[2] (D)

Answers may vary. Sample answer:

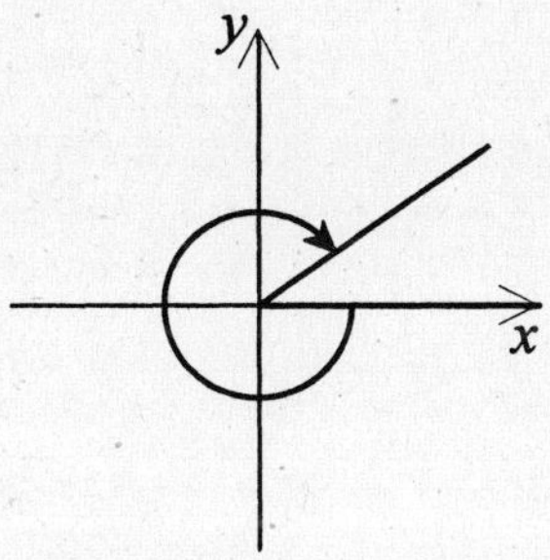

[3]

Answers may vary. Sample answers:

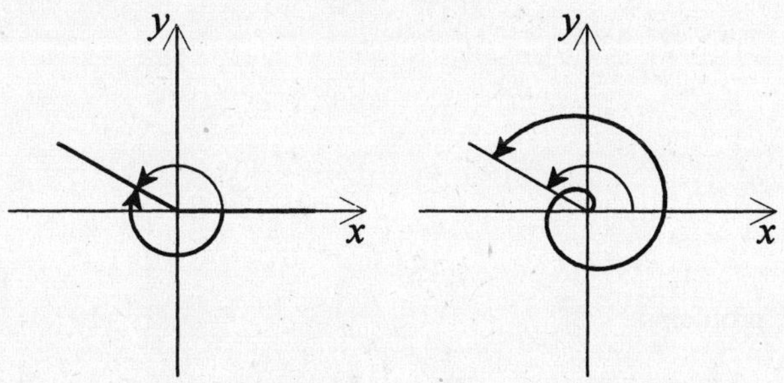

[4]

Objective 2: Use radian measure

[5] (B)

[6] (A)

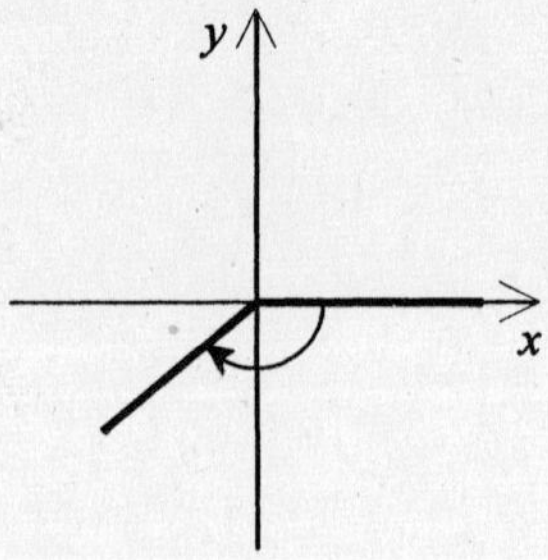

[7] ______________________________

Positive coterminal angle: $\frac{5\pi}{3}$

Negative coterminal angle: $-\frac{7\pi}{3}$

[8] ______________________________

Objective 3: Use degree measure

[9] (B)

[10] (D)

[11] 100°

[12] 0.925

Objective 4: Use angles to model and solve real-life problems

[13] (A)

[14] (D)

[15] 0.8 ft/min

[16] 8° 31′ 40″

Section 1.2: Trigonometric Functions: The Unit Circle

Objective 1: Identify a unit circle and its relationship to real numbers

[17] (C)

[18] (D)

[19] 0.70 ft

[20] 0.982

Objective 2: Evaluate trigonometric functions using the unit circle

[21] (A)

[22] (A)

[23] $\sin t = \frac{\sqrt{2}}{2}$, $\csc t = \sqrt{2}$
$\cos t = \frac{\sqrt{2}}{2}$, $\sec t = \sqrt{2}$
$\tan t = 1$, $\cot t = 1$

[24] $\frac{1}{x}$

Objective 3: Use the domain and period to evaluate sine and cosine functions

[25] (A)

[26] (B)

[27] Answers may vary. Sample answer: $\cos\frac{33\pi}{4} = \cos\left(8\pi + \frac{\pi}{4}\right) = \cos\frac{\pi}{4} = \frac{\sqrt{2}}{2}$.

[28] odd function

Objective 4: Use a calculator to evaluate trigonometric functions

[29] (D)

[30] (C)

[31] $\cot t = 1.8305$; $\sec t = 1.1395$; $\csc t = 2.0858$

[32] $\sin t = 0.3420$; $\cos t = -0.9397$; $\tan t = -0.3640$

Section 1.3: Right Triangle Trigonometry

Objective 1: Evaluate trigonometric functions of acute angles

[33] (C)

[34] (C)

[35] $\sin\theta = \frac{1}{2}$; $\cos\theta = \frac{\sqrt{3}}{2}$; $\tan\theta = \frac{\sqrt{3}}{3}$; $\cot\theta = \sqrt{3}$; $\sec\theta = \frac{2\sqrt{3}}{3}$; $\csc\theta = 2$

[36] Answers may vary. Sample answer: $\sin\theta = \frac{1}{2}$.

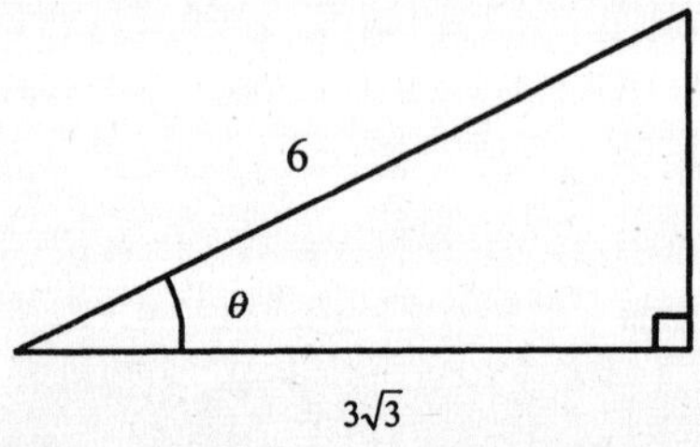

Objective 2: Use the fundamental trigonometric identities

[37] (A)

[38] (D)

[39] Answers may vary. Sample answer: $-\tan^2\beta$

[40] $\frac{2\sqrt{6}}{5}$

Objective 3: Use a calculator to evaluate trigonometric functions

[41] (A)

[42] (D)

[43] 2.2642

[44] 1.7893

Objective 4: Use trigonometric functions to model and solve real-life problems

[45] (D)

[46] (A)

[47] 49.51 m

[48] 8 ft

Section 1.4: Trigonometric Functions of Any Angle

Objective 1: Evaluate trigonometric functions of any angle

[49] (A)

[50] (B)

[51] $\sin\frac{3\pi}{2} = -1,\ \cos\frac{3\pi}{2} = 0,\ \tan\frac{3\pi}{2} = \text{undefined}$

[52] $\cot\theta = -\frac{7}{24},\ \sec\theta = -\frac{25}{7},\ \csc\theta = \frac{25}{24}$

Objective 2: Use reference angles to evaluate trigonometric functions

[53] (D)

[54] (C)

[55] $-\sin 58°$

[56] $\tan\theta = -\tan\theta' = -0.9004$

Objective 3: Evaluate trigonometric functions of real numbers

[57] (D)

[58] (B)

[59] $\frac{\sqrt{3}}{3}$

[60] $\sin\theta = -\frac{\sqrt{3}}{2}$

Section 1.5: Graphs of Sine and Cosine Functions

Objective 1: Sketch the graphs of basic sine and cosine functions

[61] (C)

[62] (D)

[63] $(0,\ 2.5),\ \left(\frac{\pi}{2},\ 0\right),\ (\pi,\ -2.5),\ \left(\frac{3\pi}{2},\ 0\right)$

[64] 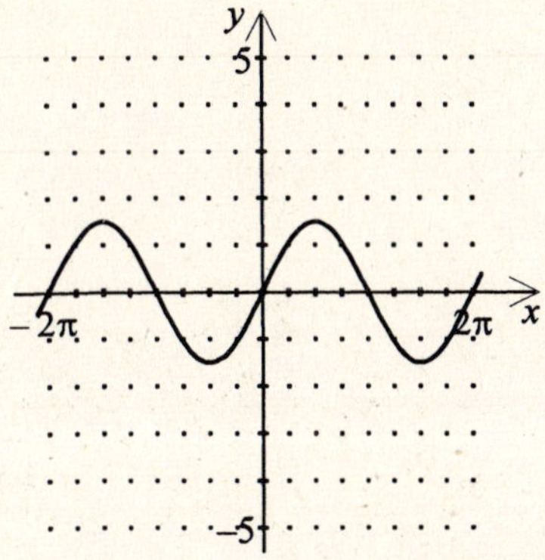

Objective 2: Use amplitude and period to help sketch the graphs of sine and cosine functions

[65] (A)

[66] (A)

[67] Answers may vary. Sample answer:

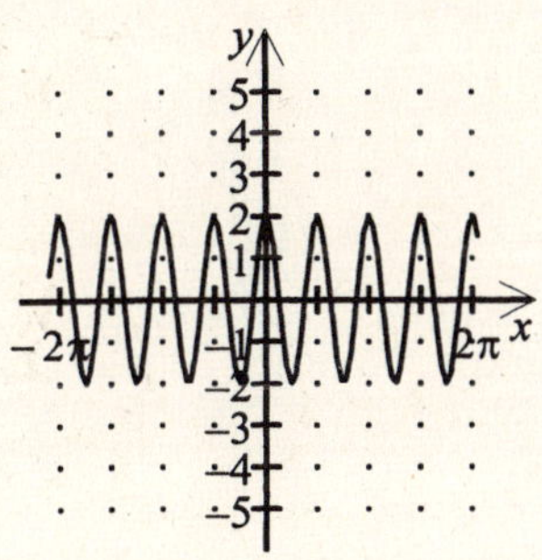

[68] Amplitude = 3; period = $\frac{3\pi}{4}$

Objective 3: Sketch translations of the graphs of sine and cosine functions

[69] (C)

[70] (B)

[71] Answers may vary. Sample answer: $\left(-\frac{5\pi}{6},\ 0\right), \left(-\frac{\pi}{2},\ 0\right)$

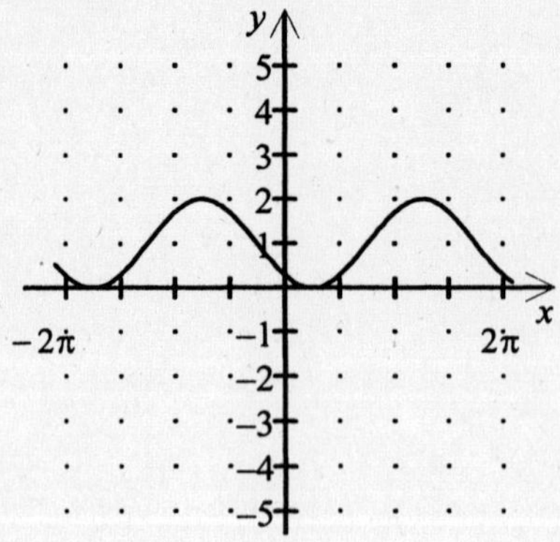

[72] ____________________

Objective 4: Use sine and cosine functions to model real-life data

[73] (B)

[74] (D)

(a) $h(t) = 57 + 31 \sin\left(\frac{5\pi t}{3}\right)$

[75] (b) 30.2 centimeters

[76] $h(t) = 28\cos(0.491t - 1.473) + 80$

Section 1.6: Graphs of Other Trigonometric Functions

Objective 1: Sketch the graphs of tangent functions

[77] (D)

[78] (C)

Answers may vary. Sample answer: $x = -\dfrac{\pi}{36}$, $x = -\dfrac{17\pi}{36}$

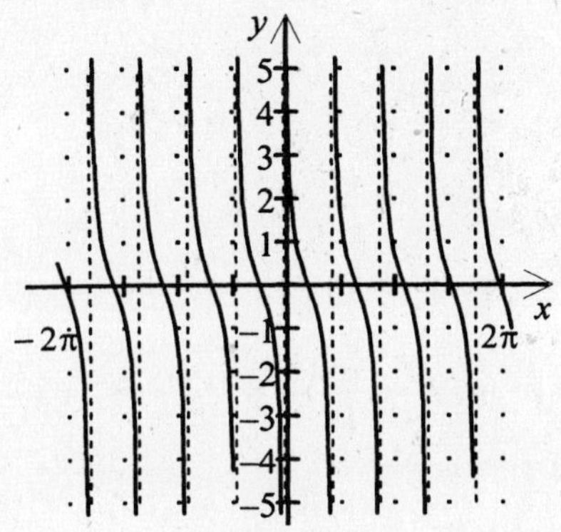

[79]

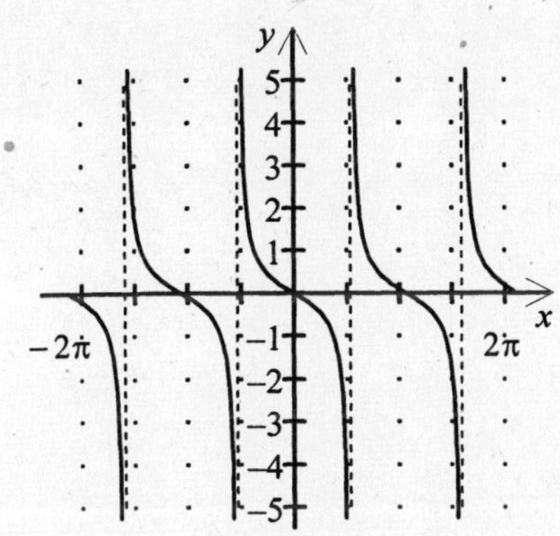

[80]

Objective 2: Sketch the graphs of cotangent functions

[81] (A)

[82] (A)

Answers may vary. Sample answer: $x = -\frac{\pi}{2}$, $x = -\frac{\pi}{18}$.

[83] 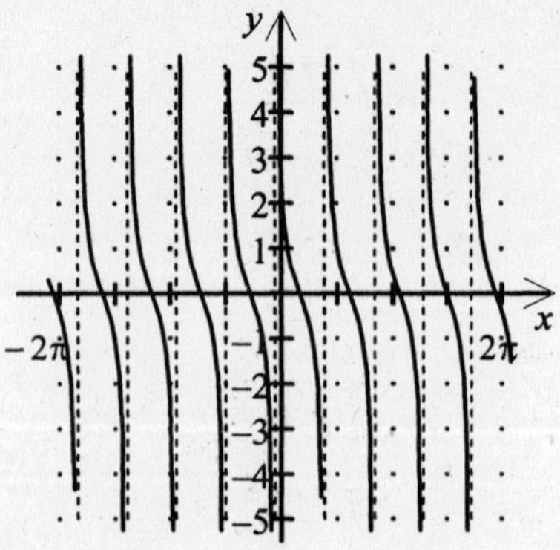

[84] 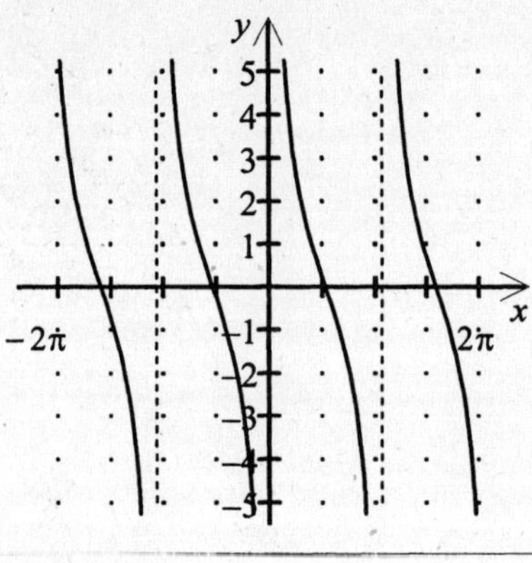

Objective 3: Sketch the graphs of secant and cosecant functions

[85] (D)

[86] (D)

Answers may vary. Sample answer: $x = \dfrac{\pi}{20}$, $x = -\dfrac{11\pi}{20}$

[87] 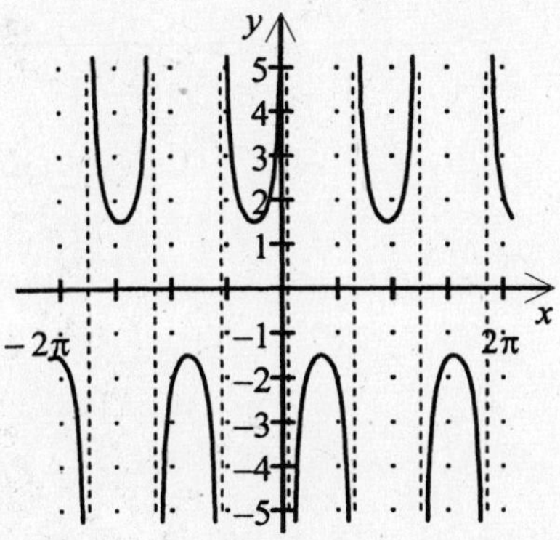

[88] period = 6π, phase shift = $-\pi$

Objective 4: Sketch the graphs of damped trigonometric functions

[89] (D)

[90] (A)

[91] $\left(\dfrac{e}{2}\right)^{-x/4}$

[92] 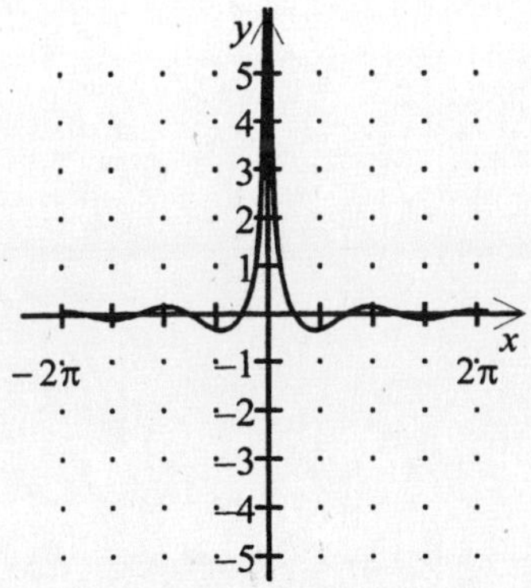

Section 1.7: Inverse Trigonometric Functions

Objective 1: Evaluate the inverse sine function

[93] (B)

[94] (D)

[95] $\frac{\pi}{4}$

[96] $\theta = 0$

Objective 2: Evaluate the other inverse trigonometric functions

[97] (B)

[98] (B)

[99] $\theta = \frac{\pi}{6}$

[100] $\left(-\frac{\sqrt{3}}{2}\right)$

Objective 3: Evaluate the compositions of trigonometric functions

[101] (A)

[102] (A)

[103] $\frac{3\pi}{8}$

[104] $\frac{1}{2}$

Section 1.8: Applications and Models

Objective 1: Solve real-life problems involving right triangles

[105] (A)

[106] (B)

[107] roof line = 9.8 ft long, shed depth = 9.0 ft

[108] 2250 ft

Objective 2: Solve real-life problems involving directional bearings

[109] (C)

[110] (D)

[111] N 39.9° E , distance = 577.2 mi

[112] N 22° E

Objective 3: Solve real-life problems involving harmonic motion

[113] (D)

[114] (B)

[115] $d(t) = 6.3 \sin \frac{2\pi}{15} t$

[116] $x(t) = 5 \cos 3.31\, t,\ x = -2.04$ cm

Chapter 2 Analytic Trigonometry

Section 2.1: Using Fundamental Identities

Objective 1: Recognize and write the fundamental trigonometric identities

1. Which expression completes the fundamental trigonometric identity? $\csc(-x) =$

(A) $\sin x$ (B) $\csc x$ (C) $-\sin x$ (D) $-\csc x$

2. Which is a trigonometric identity?

(A) $\sec u = \dfrac{1}{\cos u}$ (B) $\cos\left(\dfrac{\pi}{2} + u\right) = -\cos u$ (C) $\cot u = -\dfrac{\sin u}{\cos u}$ (D) $\tan^2 u + \cot^2 u = 1$

3. Is the equation a valid form of one of the Pythagorean trigonometric identities? $1 + \cot^2 \beta = \csc^2 \beta$

4. Write the radical form of one of the Pythagorean trigonometric identities.

Objective 2: Use the fundamental trigonometric identities to evaluate trigonometric functions, simplify trigonometric expressions, and rewrite trigonometric expressions

5. Factor the expression and use the fundamental identities to simplify.

$\cos^2 x \csc^2 x - \cos^2 x$

(A) $\cos^2 x$ (B) 1 (C) $\cos^2 x \cot^2 x$ (D) $\cos^2 x \tan^2 x$

6. Use trigonometric substitution to write the algebraic expression as a trigonometric function of θ, where $0 < \theta < \dfrac{\pi}{2}$.

$\sqrt{36 - 4x^2}$, $x = 3\cos\theta$

(A) $6 + 3\cos\theta$ (B) $12\sin\theta$ (C) $36\sin^2\theta$ (D) $6\sin\theta$

7. Use the fundamental identities to write the expression in terms of a single trigonometric function.

$\dfrac{\sec x}{\cos x} - \sec x \cos x$

8. Perform the addition and use the fundamental identities to simplify.

$\dfrac{4 + 2\cos x}{\sin x} + \dfrac{2\sin x}{1 + \cos x}$

Section 2.2: Verifying Trigonometric Identities

Objective 1: Plan a strategy for verifying trigonometric identities

9. Convert all of the terms to sines and cosines and simplify to find the expression that completes the identity.

$$\frac{\csc x}{\tan x} =$$

(A) $\sin x$ (B) $\frac{1}{\sin x}$ (C) $\frac{\sin^2 x}{\cos x}$ (D) $\frac{\cos x}{\sin^2 x}$

10. Add or subtract the fractions. Then simplify using the Pythagorean identities and factoring to find the expression that completes the identity.

$$\frac{\cot x}{\csc x + 1} + \frac{\csc x + 1}{\cot x}$$

(A) $2 \csc x$ (B) $2 \cot x$ (C) $2 \sec x$ (D) $2 \tan x$

11. Which fundamental identities could be used to verify the identity?

$$\tan\theta\cos\theta = \sin\theta$$

12. Convert all of the terms to sines and cosines and simplify.

$$\frac{\sec x}{\cot x}$$

Objective 2: Verify trigonometric identities

Identify the expression that completes the equation so that it is an identity.

13. $\frac{1 + \sin u}{\cos u} + \frac{\cos u}{1 + \sin u} =$ (A) $-2 \sin u$ (B) $2 \sec u$ (C) 0 (D) $2 + \cos u$

14. $\frac{\cos x}{1 - \sin x} =$ (A) $\frac{1 + \csc x}{\csc x}$ (B) $\sin x$ (C) $\frac{1 + \sin x}{\cos x}$ (D) $\frac{1 + \sec x}{\sec x}$

Verify the identity.

15. $\cot x \left(\cos x \tan x + \sin x\right) = 2 \cos x$

Verify the identity.

16. $\dfrac{\cos x}{1-\sin x} = \sec x + \tan x$

Section 2.3: Solving Trigonometric Equations

Objective 1: Use standard algebraic techniques to solve trigonometric equations

Identify the x-values that are solutions of the equation.

17. $4\sqrt{3}\tan x - 9 = 13\sqrt{3}\tan x$

(A) $x = \dfrac{\pi}{6}, \dfrac{7\pi}{6}$ (B) $x = \dfrac{7\pi}{6}, \dfrac{11\pi}{6}$ (C) $x = \dfrac{\pi}{6}, \dfrac{11\pi}{6}$ (D) $x = \dfrac{5\pi}{6}, \dfrac{11\pi}{6}$

18. $18\cot^2 x - 6 = 0$

(A) $\dfrac{\pi}{3}, \dfrac{2\pi}{3}, \dfrac{4\pi}{3}, \dfrac{5\pi}{3}$ (B) $\dfrac{\pi}{4}, \dfrac{3\pi}{4}, \dfrac{5\pi}{4}, \dfrac{7\pi}{4}$

(C) $\dfrac{\pi}{6}, \dfrac{5\pi}{6}, \dfrac{7\pi}{6}, \dfrac{11\pi}{6}$ (D) $\dfrac{\pi}{4}, \dfrac{\pi}{2}, \dfrac{2\pi}{3}, \dfrac{5\pi}{6}$

19. Solve the equation.
$5\csc x - 12 = 11\csc x$

20. Find all solutions of the equation in the interval $[0, 2\pi)$.
$2\cos x\sin x + \sin x = 0$

Objective 2: Solve trigonometric equations of quadratic type

Find all solutions of the equation in the interval $[0, 2\pi)$.

21. $\tan^2\theta = \dfrac{\sqrt{2}}{2}\sec\theta$ (A) $\dfrac{\pi}{4}, \dfrac{7\pi}{4}$ (B) $\dfrac{3\pi}{4}, \dfrac{5\pi}{4}$ (C) $\dfrac{5\pi}{6}, \dfrac{7\pi}{6}$ (D) None of these

Find all solutions of the equation in the interval $[0, 2\pi)$.

22. $2\cos^2 x = 13\sin x - 5$

(A) $0, \frac{3\pi}{4}, \pi, \frac{7\pi}{4}$ (B) $\frac{\pi}{6}, \frac{5\pi}{6}$ (C) $\frac{\pi}{6}, \frac{\pi}{2}, \frac{5\pi}{6}, \frac{3\pi}{2}$ (D) $\frac{3\pi}{4}, \frac{7\pi}{4}$

23. $-\sin x + 1 = 2\cos^2 x$

24. $2\tan^2 2x = 3\sec 2x$

Objective 3: Solve trigonometric equations involving multiple angles

Find all solutions of the equation in the interval $[0, 2\pi)$.

25. $3\sec 3x - 2\sqrt{3} = 0$

(A) $\frac{\pi}{18}, \frac{11\pi}{18}, \frac{13\pi}{18}, \frac{23\pi}{18}, \frac{25\pi}{18}$ (B) $\frac{\pi}{18}, \frac{13\pi}{18}$

(C) $\frac{\pi}{18}, \frac{11\pi}{18}, \frac{2\pi}{3}, \frac{13\pi}{18}, \frac{23\pi}{18}, \frac{25\pi}{18}, \frac{35\pi}{18}$ (D) $\frac{\pi}{18}, \frac{11\pi}{18}, \frac{13\pi}{18}, \frac{23\pi}{18}, \frac{25\pi}{18}, \frac{35\pi}{18}$

26. $4\sec^2\frac{x}{2} - 5\sec\frac{x}{2} - 6 = 0$ (A) $\frac{4\pi}{3}$ (B) $\frac{2\pi}{3}$ (C) 0 (D) 2π

27. Find all solutions of the equation.

$\sin 4x = \frac{1}{2}$

28. Find all solutions of the equation in the interval $[0, 2\pi)$.

$3\tan^2 6x - 1 = 0$

Objective 4: Use inverse trigonometric functions to solve trigonometric equations

Use inverse functions where needed to find all solutions of the equation in the interval $[0,\ 2\pi)$.

29. $6\sin^2 x + \left(-2 - 3\sqrt{2}\right)\cos x - \sqrt{2} - 6 = 0$

(A) $\arccos(3),\ \arccos(3) + \pi,\ \dfrac{3\pi}{4},\ \dfrac{5\pi}{4}$ (B) $\arccos(3),\ \arccos(3) + \pi,\ \dfrac{\pi}{4},\ \dfrac{7\pi}{4}$

(C) $\arccos\left(-\dfrac{1}{3}\right),\ 2\pi - \arccos\left(-\dfrac{1}{3}\right),\ \dfrac{\pi}{4},\ \dfrac{7\pi}{4}$ (D) $\arccos\left(-\dfrac{1}{3}\right),\ 2\pi - \arccos\left(-\dfrac{1}{3}\right),\ \dfrac{3\pi}{4},\ \dfrac{5\pi}{4}$

30. $27\csc^2 x + \left(3 + 9\sqrt{3}\right)\cot x + \sqrt{3} - 27 = 0$

(A) $\operatorname{arccot}(9),\ \operatorname{arccot}(9) + \pi,\ \dfrac{2\pi}{3},\ \dfrac{5\pi}{3}$ (B) $\operatorname{arccot}(9),\ \operatorname{arccot}(9) + \pi,\ \dfrac{\pi}{3},\ \dfrac{4\pi}{3}$

(C) $\operatorname{arccot}\left(-\dfrac{1}{9}\right),\ \operatorname{arccot}\left(-\dfrac{1}{9}\right) + \pi,\ \dfrac{\pi}{3},\ \dfrac{4\pi}{3}$ (D) $\operatorname{arccot}\left(-\dfrac{1}{9}\right),\ \operatorname{arccot}\left(-\dfrac{1}{9}\right) + \pi,\ \dfrac{2\pi}{3},\ \dfrac{5\pi}{3}$

Use inverse functions where needed to find all solutions of the equation in the interval $[0,\ 2\pi)$.

31. $6\sin^2 x + \left(-2 + 3\sqrt{2}\right)\cos x + \sqrt{2} - 6 = 0$

32. $5\sqrt{3}\tan^2 x + \left(10 - 8\sqrt{3}\right)\sec x - 16 + 5\sqrt{3} = 0$

Section 2.4: Sum and Difference Formulas

Objective 1: Use sum and difference formulas to evaluate trigonometric functions, verify identities, and solve trigonometric equations

33. Find the exact value of $\cos(255°)$.

(A) $\dfrac{-\sqrt{2} + \sqrt{6}}{4}$ (B) $\dfrac{-\sqrt{6} + \sqrt{2}}{4}$ (C) $\dfrac{\sqrt{6} + \sqrt{2}}{4}$ (D) $\dfrac{-\sqrt{2} - \sqrt{6}}{4}$

34. Identify the expression that completes the identity.

$\cos\left(\frac{3\pi}{4}+x\right)=$

(A) $-\frac{\sqrt{2}}{2}(\cos x+\sin x)$

(B) $\frac{\sqrt{2}}{2}(\sin x-\cos x)$

(C) $\frac{\sqrt{2}}{2}(\cos x+\sin x)$

(D) $\frac{\sqrt{2}}{2}(\cos x-\sin x)$

35. Find the exact value of the expression.

$\cos\frac{\pi}{12}\cos\frac{\pi}{4}-\sin\frac{\pi}{12}\sin\frac{\pi}{4}$

36. Verify the identity.

$\sin\left(x-\frac{3\pi}{2}\right)=\cos x$

Section 2.5: Multiple-Angle and Product-to-Sum Formulas

Objective 1: Use multiple-angle formulas to rewrite and evaluate trigonometric functions

37. Find the exact value of $\tan 2x$ using the double angle formula.

$\tan x=\frac{2}{9},\quad 0<x<\frac{\pi}{2}$

(A) $-\frac{4\sqrt{77}}{81}$

(B) $\frac{4\sqrt{77}}{81}$

(C) $\frac{36}{77}$

(D) $\frac{73}{81}$

38. Find the exact solutions to the equation in the interval $[0,\ 2\pi)$.

$-2\sin 4x=-4\cos 2x$

(A) $\frac{\pi}{4},\frac{3\pi}{4},\frac{5\pi}{4},\frac{7\pi}{4}$

(B) $\frac{\pi}{6},\frac{\pi}{8},\frac{5\pi}{6},\frac{3\pi}{4},\frac{11\pi}{12},\frac{5\pi}{4},\frac{7\pi}{4}$

(C) $\frac{\pi}{8},\frac{\pi}{4},\frac{5\pi}{12},\frac{3\pi}{4},\frac{13\pi}{12},\frac{5\pi}{4},\frac{7\pi}{4}$

(D) $\frac{\pi}{6},\frac{5\pi}{6},\frac{7\pi}{6},2\pi$

39. Use the figure to find the exact value of the trigonometric function.
$\cos 2\theta$

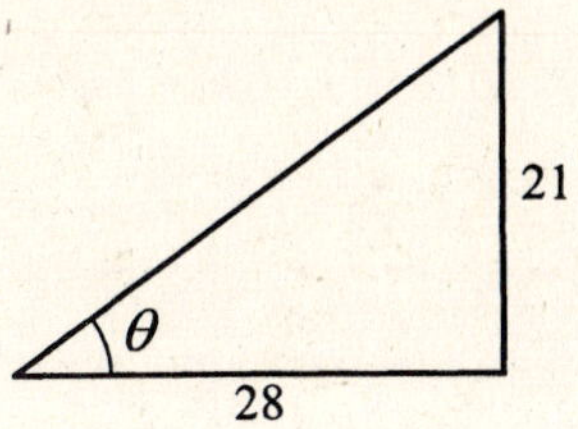

40. Use a double angle formula to rewrite the expression.
$8 \sin x \cos x$

Objective 2: Use power-reducing formulas to rewrite and evaluate trigonometric functions

Use the power-reducing formulas to find the exact value of the trigonometric function.

41. $\sin^2 15°$ (A) $\frac{3+\sqrt{2}}{6}$ (B) $\frac{2-\sqrt{3}}{4}$ (C) $\frac{3-\sqrt{2}}{6}$ (D) $\frac{2+\sqrt{3}}{4}$

42. $\cos^2 \frac{\pi}{8}$ (A) $\frac{2-\sqrt{2}}{4}$ (B) $\frac{3-\sqrt{2}}{4}$ (C) $\frac{3+\sqrt{2}}{4}$ (D) $\frac{2+\sqrt{2}}{4}$

43. $\tan^2 165°$

44. $\sin^2 \frac{37\pi}{12}$

Objective 3: Use half-angle formulas to rewrite and evaluate trigonometric functions

45. Use the figure to find the exact value of the trigonometric function.

$\cos\frac{\theta}{2}$

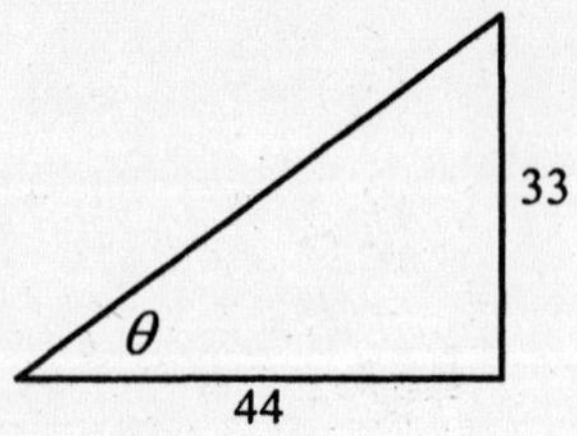

(A) $\frac{\sqrt{10}}{3}$ (B) 3 (C) $\frac{\sqrt{10}}{10}$ (D) $\frac{3\sqrt{10}}{10}$

46. Use the half-angle formulas to simplify the expression.

$\sqrt{\frac{1+\cos 8x}{1-\cos 8x}}$

(A) $-\tan 4x$ (B) $-\cot 4x$ (C) $|\cot 4x|$ (D) $|\tan 4x|$

47. Find all solutions in the interval $[0,\ 2\pi)$.

$2\sin^2\frac{x}{2} - 2\cos^2 x = 0$

48. Find the exact value of cos 105°.

Objective 4: Use product-to-sum and sum-to-product formulas to rewrite and evaluate trigonometric functions

49. Express $\sin 7x - \sin 5x$ as a product containing only sines and/or cosines.

(A) $2\sin x\cos 6x$ (B) $-2\sin x\cos 6x$ (C) $2\sin 6x\cos x$ (D) $-2\sin 6x\cos x$

Find all solutions in the interval $[0, 2\pi)$.

50. $\sin 5x + \sin x = 0$

(A) $0, \frac{\pi}{6}, \frac{\pi}{2}, \frac{5\pi}{6}, \pi, \frac{7\pi}{6}, \frac{3\pi}{2}, \frac{11\pi}{6}$

(B) $0, \frac{\pi}{3}, \frac{\pi}{2}, \frac{2\pi}{3}, \pi, \frac{3\pi}{2}$

(C) $\frac{\pi}{6}, \frac{\pi}{4}, \frac{\pi}{2}, \frac{3\pi}{4}, \frac{5\pi}{6}, \frac{7\pi}{6}, \frac{5\pi}{4}, \frac{3\pi}{2}, \frac{7\pi}{4}, \frac{11\pi}{6}$

(D) $0, \frac{\pi}{4}, \frac{\pi}{3}, \frac{2\pi}{3}, \frac{3\pi}{4}, \pi, \frac{5\pi}{4}, \frac{4\pi}{3}, \frac{5\pi}{3}, \frac{7\pi}{4}$

51. $\sin^2 3x - \sin^2 x = 0$

52. Express $\cos 8\theta \cos \theta$ as a sum containing only sines or cosines.

Answer Key for Chapter 2 Analytic Trigonometry

Section 2.1: Using Fundamental Identities

Objective 1: Recognize and write the fundamental trigonometric identities

[1] (D)

[2] (A)

[3] Yes

[4] Answers may vary. Sample answer: $\cot u = \pm\sqrt{\csc^2 u - 1}$

Objective 2: Use the fundamental trigonometric identities to evaluate trigonometric functions, simplify trigonometric expressions, and rewrite trigonometric expressions

[5] (C)

[6] (D)

[7] $\tan^2 x$

[8] Answers may vary. Sample answer: $\dfrac{6}{\sin x}$

Section 2.2: Verifying Trigonometric Identities

Objective 1: Plan a strategy for verifying trigonometric identities

[9] (D)

[10] (C)

[11] Quotient identity

[12] $\dfrac{\sin x}{\cos^2 x}$

Objective 2: Verify trigonometric identities

[13] (B)

[14] (C)

[15] Answers will vary.

[16] Answers will vary.

Section 2.3: Solving Trigonometric Equations

Objective 1: Use standard algebraic techniques to solve trigonometric equations

[17] (D)

[18] (A)

[19] $x = \frac{7\pi}{6} + 2n\pi, \frac{11\pi}{6} + 2n\pi$

[20] $0, \frac{2\pi}{3}, \pi, \frac{4\pi}{3}$

Objective 2: Solve trigonometric equations of quadratic type

[21] (A)

[22] (B)

[23] $x = \frac{7\pi}{6}, x = \frac{11\pi}{6}, x = \frac{\pi}{2}$

[24] $\frac{\pi}{6}, \frac{5\pi}{6}, \frac{7\pi}{6}, \frac{11\pi}{6}$

Objective 3: Solve trigonometric equations involving multiple angles

[25] (D)

[26] (B)

[27] $\frac{\pi}{24}+\frac{n}{2}\pi,\ \frac{5\pi}{24}+\frac{n}{2}\pi$

[28] $\frac{\pi}{36}, \frac{5\pi}{36}, \frac{7\pi}{36}, \frac{11\pi}{36}, \frac{13\pi}{36}, \frac{17\pi}{36}, \frac{19\pi}{36}, \frac{23\pi}{36}, \frac{25\pi}{36}, \frac{29\pi}{36}, \frac{31\pi}{36}, \frac{35\pi}{36}, \frac{37\pi}{36}, \frac{41\pi}{36}, \frac{43\pi}{36}, \frac{47\pi}{36}, \frac{49\pi}{36}, \frac{53\pi}{36}, \frac{55\pi}{36}, \frac{59\pi}{36}, \frac{61\pi}{36}, \frac{65\pi}{36}, \frac{67\pi}{36}, \frac{71\pi}{36}$

Objective 4: Use inverse trigonometric functions to solve trigonometric equations

[29] (D)

[30] (D)

[31] $\arccos\left(-\frac{1}{3}\right),\ 2\pi-\arccos\left(-\frac{1}{3}\right),\ \frac{\pi}{4},\ \frac{7\pi}{4}$

[32] $\operatorname{arcsec}\left(\frac{8}{5}\right), 2\pi-\operatorname{arcsec}\left(\frac{8}{5}\right), \frac{5\pi}{6}, \frac{7\pi}{6}$

Section 2.4: Sum and Difference Formulas

Objective 1: Use sum and difference formulas to evaluate trigonometric functions, verify identities, and solve trigonometric equations

[33] (B)

[34] (A)

[35] $\frac{1}{2}$

[36] Answers will vary.

Section 2.5: Multiple-Angle and Product-to-Sum Formulas

Objective 1: Use multiple-angle formulas to rewrite and evaluate trigonometric functions

[37] (C)

[38] (A)

[39] $\cos 2\theta = \frac{7}{25}$

[40] $4 \sin 2x$

Objective 2: Use power-reducing formulas to rewrite and evaluate trigonometric functions

[41] (B)

[42] (D)

[43] $7 - 4\sqrt{3}$

[44] $\frac{2-\sqrt{3}}{4}$

Objective 3: Use half-angle formulas to rewrite and evaluate trigonometric functions

[45] (D)

[46] (C)

[47] $\frac{1}{3}\pi,\ \pi,\ \frac{5}{3}\pi$

[48] $-\frac{\sqrt{2-\sqrt{3}}}{2}$

Objective 4: Use product-to-sum and sum-to-product formulas to rewrite and evaluate trigonometric functions

[49] (A)

[50] (D)

[51] $0, \frac{\pi}{4}, \frac{\pi}{2}, \frac{3\pi}{4}, \pi, \frac{5\pi}{4}, \frac{3\pi}{2}, \frac{7\pi}{4}$

[52] $\frac{1}{2}(\cos 9\theta + \cos 7\theta)$

Chapter 3 Additional Topics in Trigonometry

Section 3.1: Law of Sines

Objective 1: Use the Law of Sines to solve oblique triangles (AAS or ASA)

1. Use the information to solve the triangle.
$B = 20°$, $C = 13°$, and $a = 5$

(A) $A = 147°$, $b \approx 31.8$, $c \approx 12.1$
(B) $A = 157°$, $b \approx 4.4$, $c \approx 2.1$
(C) $A = 157°$, $b \approx 31.8$, $c \approx 8.0$
(D) $A = 147°$, $b \approx 3.1$, $c \approx 2.1$

2. Use the information to solve the triangle for the requested side (if possible). If two solutions exist, find both.
Find a.

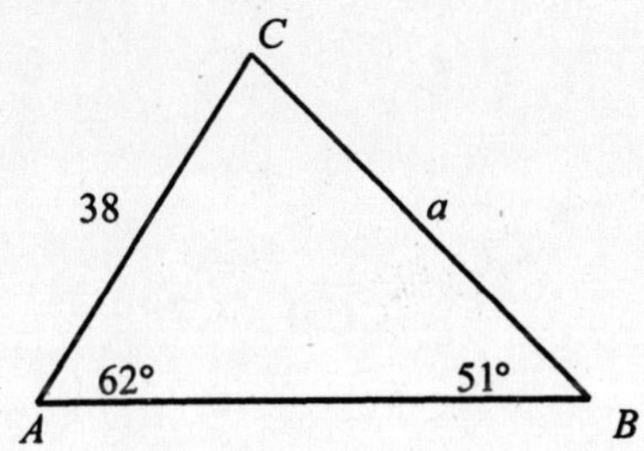

(A) $a \approx 43.2$ or $a \approx 44.2$
(B) $a \approx 42.2$
(C) $a \approx 43.2$
(D) No solution exists.

3. Use the Law of Sines to solve for b to the nearest tenth.

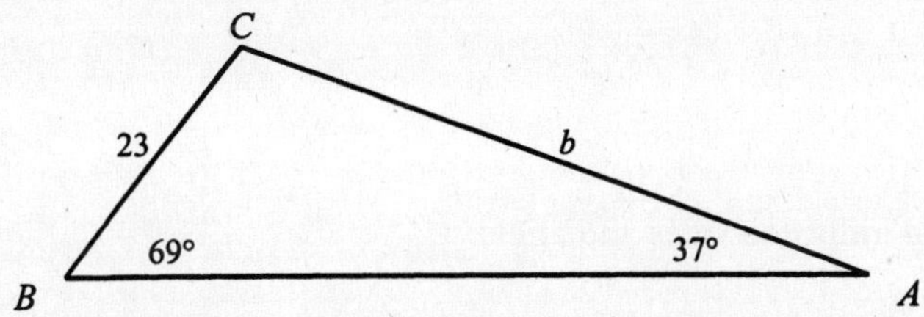

4. Given a triangle with $a = 6$, $A = 37°$, and $B = 28°$, what is the length of c? Round your answer to two decimal places.

Objective 2: Use the Law of Sines to solve oblique triangles (SSA)

5. Use the information to solve the triangle for the requested side (if possible). If two solutions exist, find both.
Find c if $A = 27°$, $a = 7$, and $b = 10$.

(A) $c \approx 14.24$ (B) $c \approx 12.21$ or $c \approx 6.09$

(C) $c \approx 14.24$ or $c \approx 3.58$ (D) No solution exists.

6. Use the Law of Sines to solve for $m\angle B$ to the nearest tenth.

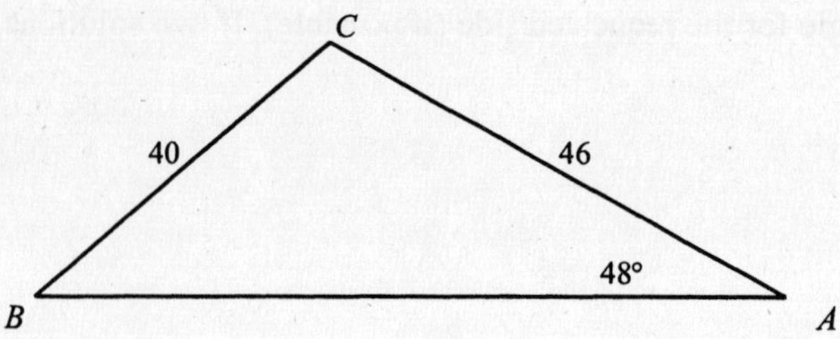

(A) 57.7 (B) 59.7 (C) 58.7 (D) 56.7

7. How many solutions exist for the triangle?
$a = 6.2$, $b = 6.5$, and $A = 83°$

8. Find a value for a such that the triangle has the number of solutions required, if possible.
One solution when $b = 15.5$ and $A = 100°$

Objective 3: Find the areas of oblique triangles

9. Find the area of the triangle having the indicated sides and angle.
$B = 59° 19'$, $a = 7$, $c = 10$

(A) 60.2 (B) 17.86 (C) 35 (D) None of these

10. Use the given measures to find the area of triangle ABC.
$A = 21°$, $a = 5.6$, $b = 5.6$

(A) 5.62 (B) 20.98 (C) 11.65 (D) 10.49

11. Triangle ABC has the given area, angle, and side. Find the length of the requested side. Round to the nearest integer.
Area = 17.46, $A = 86°$, $b = 5$. Find c.

12. Triangle ABC has the given measures. If the triangle has exactly one solution, give the area of the triangle to two decimal places. If the triangle does not have exactly one solution, determine how many solutions there are.
$A = 60°,\ a = 8,\ b = 13.3$

Objective 4: Use the Law of Sines to model and solve real-life problems

13. A loading dock ramp that is 19 feet long rises at an angle of 13.2° from the horizontal. Due to new design specifications, a longer ramp is to be used, so that the angle is reduced to 8.6°. How much farther out from the dock will that put the foot of the ramp?

(A) 29.0 ft (B) 8.4 ft (C) 35.4 ft (D) 10.2 ft

14. A pole 55 feet tall is situated at the bottom of a hill that slopes up at an angle of 12.1°. A guy wire from the top of the pole to the hillside forms an angle of 45° with the top of the pole. Find the distance from the base of the pole to the guy wire's point of attachment.

(A) 42.2 ft (B) 43.3 ft (C) 45.1 ft (D) 46.3 ft

15. A ship at sea, the Intrepid, spots two other ships, the Ranger and the Lancer, and measures the angle between them to be 42°. The distance between the Intrepid and the Ranger is 3730 meters. The Ranger measures an angle of 77° between the Intrepid and the Lancer. To the nearest meter, what is the distance between the Ranger and the Lancer?

16. An airplane left an airport and flew east for 193 miles. Then it turned northward to N 14° E. When it was 262 miles from the airport, there was an engine problem and it turned to take the shortest route back to the airport. Find θ, the angle through which the airplane turned.

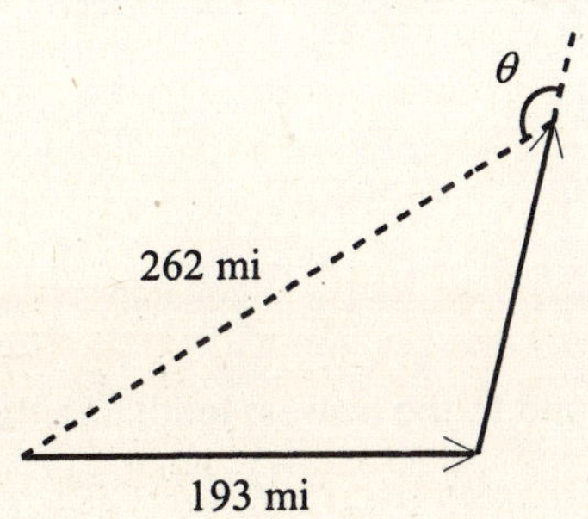

Section 3.2: Law of Cosines

Objective 1: Use the Law of Cosines to solve oblique triangles (SSS or SAS)

17. Use the given measures and the Law of Cosines to solve triangle *ABC*.
$a=12,\ b=15,\ c=14$

(A) $A=70.0°;\ B=48.7°;\ C=61.3°$ (B) $A=20.0°;\ B=118.7°;\ C=41.3°$

(C) $A=48.7°;\ B=70.0°;\ C=61.3°$ (D) $A=118.7°;\ B=20.0°;\ C=41.3°$

18. Use the Law of Cosines to find the third side of the triangle.

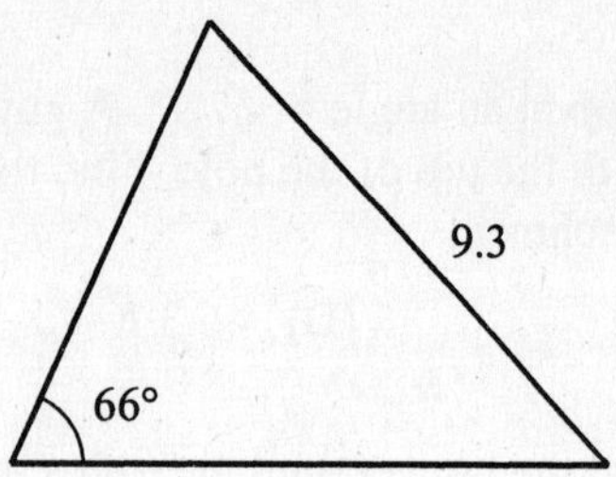

(A) 7.6 (B) 10.1 (C) 17 (D) 8.3

19. Use the Law of Cosines to find the length of the diagonal of the parallelogram.

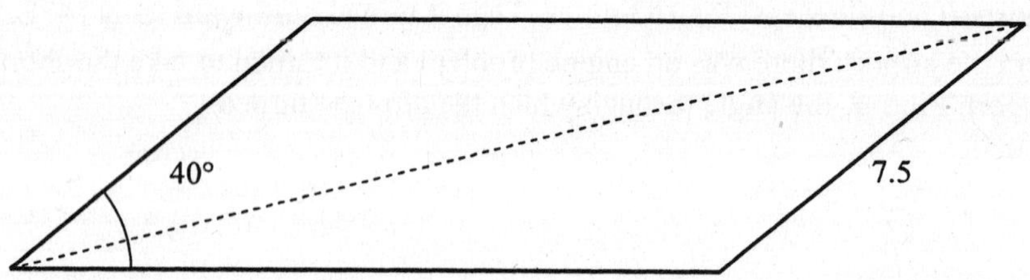

20. Use the Law of Cosines to find angle θ in the parallelogram. Round to the nearest tenth of a degree.

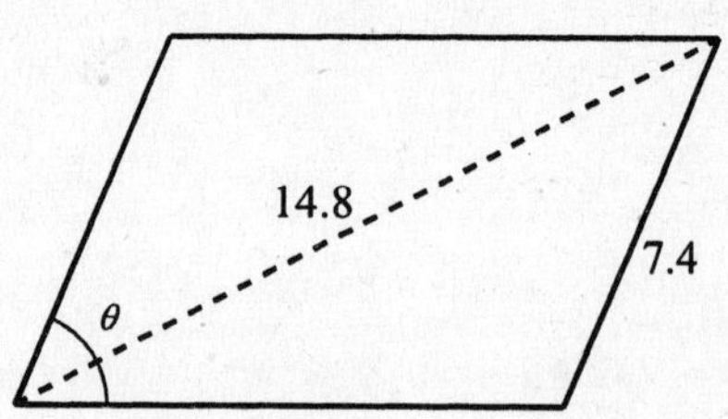

Objective 2: Use the Law of Cosines to model and solve real-life problems

21. Two ships leave a port at the same time. When ship A is 190 miles due east of the port, ship B is 145 miles from the port and 180 miles from ship A, in the direction shown below. What is ship B's bearing?

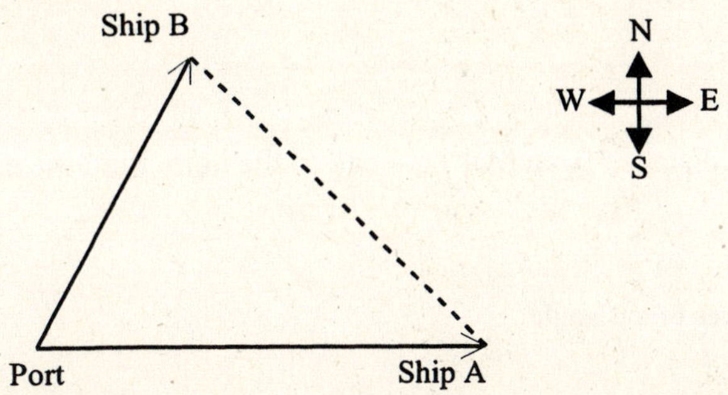

(A) N 46° W (B) N 44° W (C) N 63.3° E (D) N 26.7° E

22. A plane travels 120 miles at a heading of N 23° W. It then changes direction and travels 115 miles at a heading of N 63° W. How far is the plane from its original position?

(A) 220.8 mi (B) 231.9 mi (C) 235 mi (D) 198.8 mi

23. A ship travels due west for 82 miles. It then travels in a northern direction for 77 miles and ends up 143 miles from its original position. How many degrees did it turn when it changed direction? Round your answer to the nearest tenth.

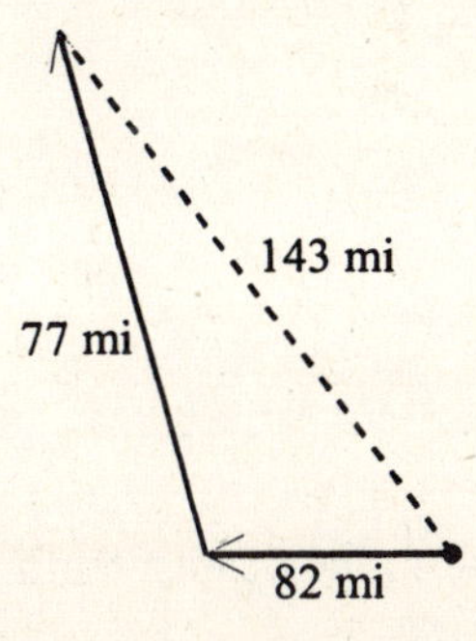

24. Martha and Erika are golfing on a beautiful summer day. The fourth hole is 340 yards from the tee. On that hole, Martha hit her drive to the left of the correct direction, as sketched below.

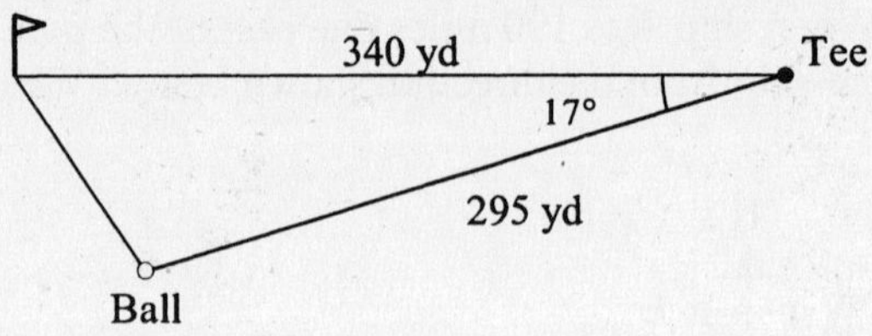

Write the Law of Cosines equation for the distance from Martha's ball to the hole. Find the distance to the nearest tenth of a yard.

Objective 3: Use Heron's Area Formula to find the area of a triangle

25. Use Heron's Area Formula to find the area of the triangle with the given measures.
Equilateral triangle, with a perimeter of 21

(A) 240.0 (B) 6.1 (C) 21.2 (D) 680.6

26. Use Heron's Area Formula to find the value of x in the triangle with the given measures.
Area = 70.2

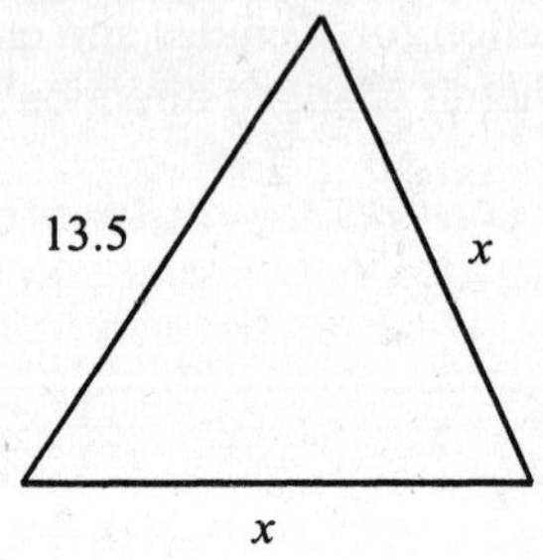

(A) $x = 11.8$ (B) $x = 12.4$ (C) $x = 16$ (D) $x = 27$

27. Use Heron's Area Formula to find the area of the triangle. Round the result to the nearest tenth.
$a = 27.5,\ b = 30,\ c = 35$

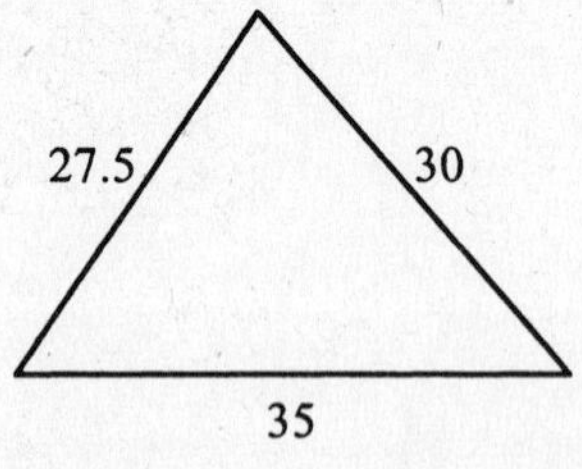

28. Use Heron's Area Formula to find the area of the triangle with the given measures.
Perimeter = 24.4

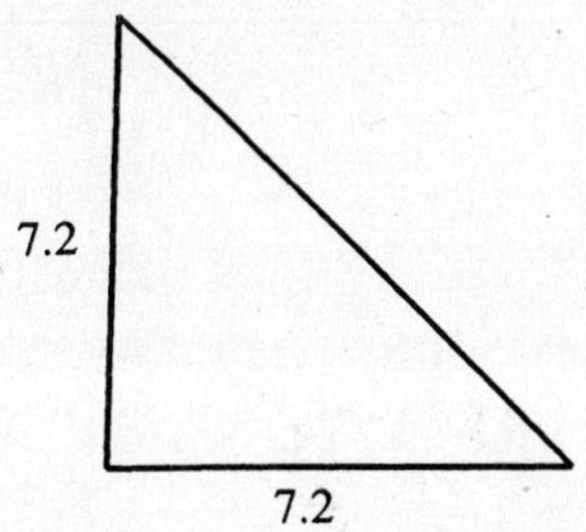

Section 3.3: Vectors in the Plane

Objective 1: Represent vectors as directed line segments

29. Identify the initial and terminal points of a vector that has the same direction as **v**.

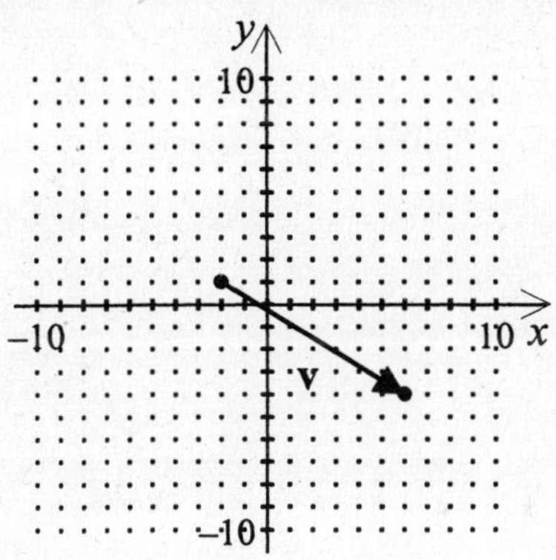

(A) $(1,\ 3)$ to $(-4,\ 11)$

(B) $(1,\ 3)$ to $(9,\ -2)$

(C) $(2,\ 1)$ to $(-6,\ -4)$

(D) $(2,\ 1)$ to $(7,\ -7)$

30. Identify the vector with the same magnitude as **v**. (A) **w** (B) **t** (C) **u** (D) **r**

$\mathbf{v} = \overrightarrow{AB}$ with $A = (-2, -4)$ and $B = (1, 0)$.

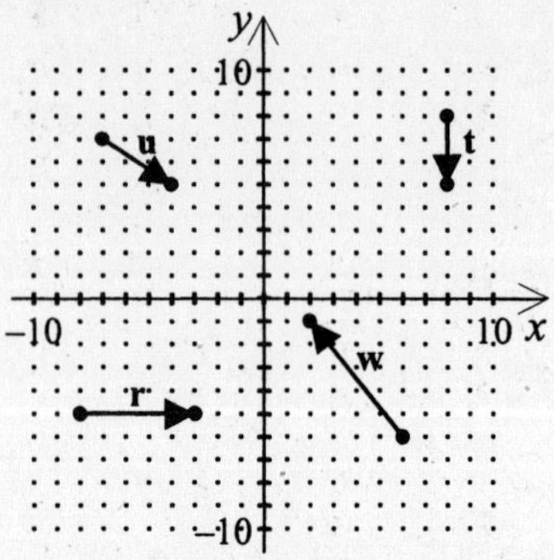

31. Show that **u** = **v**.

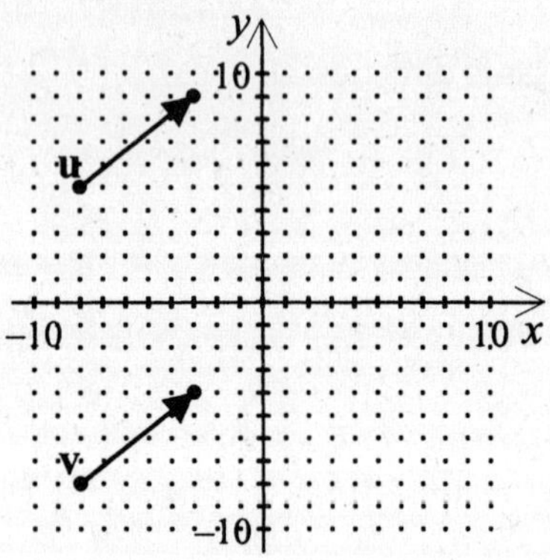

32. Find the value of k that makes $\mathbf{u} = \mathbf{v}$.

$\mathbf{u} = \overrightarrow{AB}$ with $A = (k,\ 5),\ B = (2,\ 1)$

$\mathbf{v} = \overrightarrow{MN}$ with $M = (-8,\ -6),\ N = (-3,\ -10)$

Objective 2: Write the component forms of vectors

33. Identify the pair of points that could be the initial and terminal points of the vector.

$\mathbf{u} = \langle 8, -7 \rangle$

(A) $(6,\ 9)$ and $(-1,\ 1)$ (B) $(-1,\ 1)$ and $(-9,\ -6)$

(C) $(3,\ -2)$ and $(11,\ -9)$ (D) $(-4,\ 6)$ and $(3,\ -2)$

34. Identify the initial point of vector **v**.
$\mathbf{v} = \langle -5, -5 \rangle$; terminal point is $(-2, -1)$

(A) $(8, 9)$ (B) $(3, 4)$ (C) $(4, 3)$ (D) $(-7, -6)$

35. Give the component form of the vector $\mathbf{v} = \overrightarrow{EF}$ and sketch the vector in standard position.
$E = (6, -6)$ and $F = (2, -1)$

36. Find all possible values of the missing component of the vector **u**.
$\mathbf{u} = \langle 3, y \rangle$, $\|\mathbf{u}\| = \sqrt{58}$

Objective 3: Perform basic vector operations and represent them graphically

37. Use the figure to identify the graph of the result of the specified vector operation.
$\mathbf{v} - \mathbf{w}$

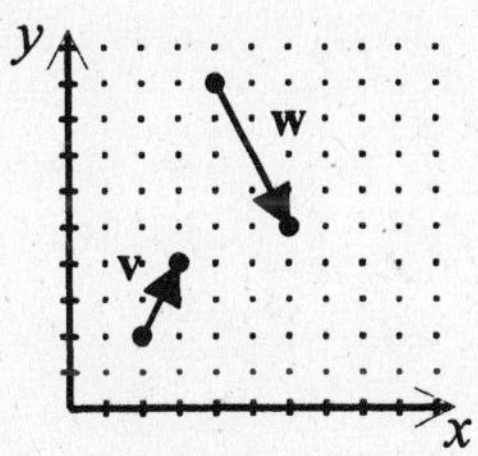

(A)
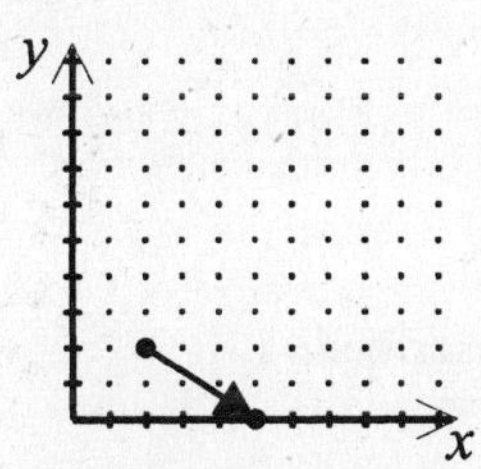

(B)
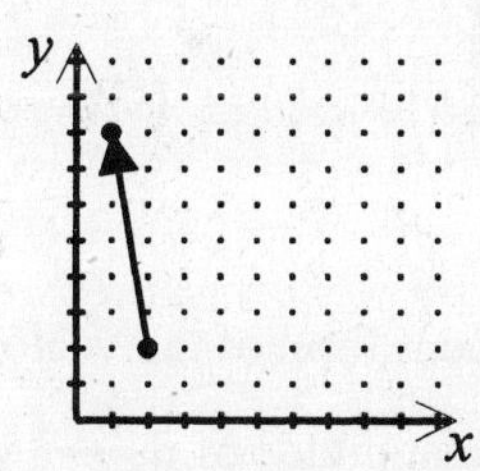

(C)
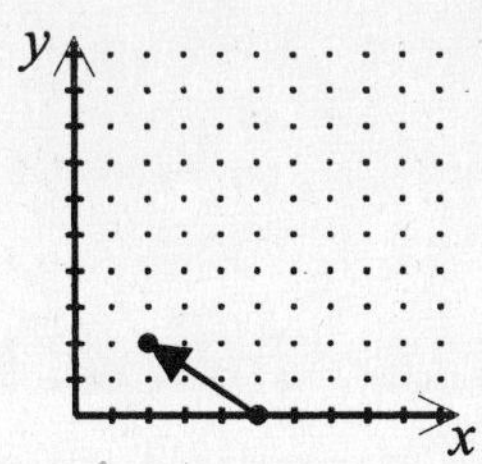

(D)
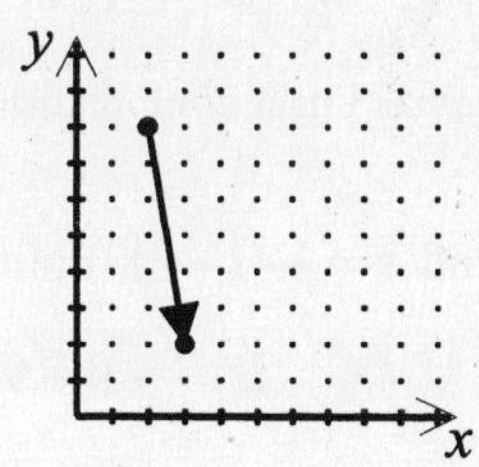

38. Identify the expression that is represented by the dashed segment in the graph.

$\mathbf{u} = \langle 1, -3 \rangle$, $\mathbf{v} = \langle 3, -2 \rangle$

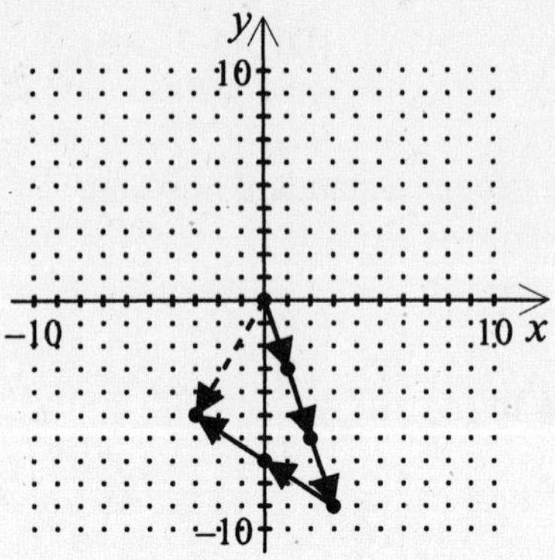

(A) $3\mathbf{v} + 2\mathbf{u}$ (B) $3\mathbf{u} - 2\mathbf{v}$ (C) $3\mathbf{v} - 2\mathbf{u}$ (D) $3\mathbf{u} + 2\mathbf{v}$

39. Find the sum of the pair of vectors. Express the answer in component form.

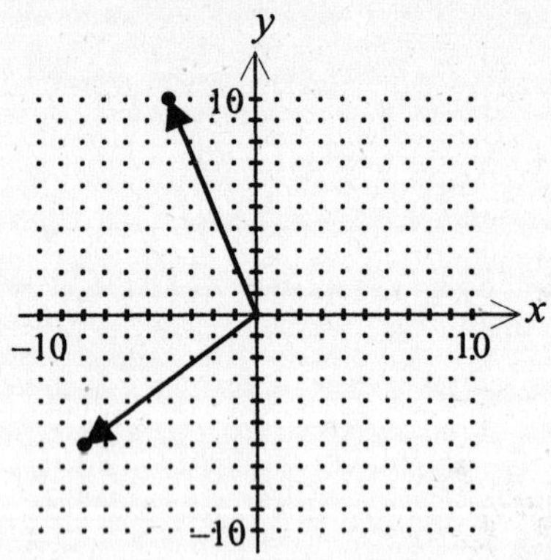

40. Give the component form of the vector **u** that has the magnitude described.

$\mathbf{v} = \langle -6, -5 \rangle$, magnitude of $\mathbf{u} = \frac{1}{7}\|\mathbf{v}\|$

Objective 4: Write vectors as linear combinations of unit vectors

41. Let $\mathbf{u} = 3\mathbf{i} - 2\mathbf{j}$ and $\mathbf{v} = -4\mathbf{i} - 3\mathbf{j}$. Find $-2\mathbf{u} - 4\mathbf{v}$.

(A) $-22\mathbf{i} + 8\mathbf{j}$ (B) $-3\mathbf{i} + 9\mathbf{j}$ (C) $10\mathbf{i} + 16\mathbf{j}$ (D) $-6\mathbf{i} + 12\mathbf{j}$

42. Find the number by which the components of the vector can be divided to find the unit vector in the same direction.
$\mathbf{u} = 6\mathbf{i} + 2\mathbf{j}$

(A) 40 (B) $2\sqrt{10}$ (C) 4 (D) 20

43. Find the unit vector in the same direction as **u**.
$\mathbf{u} = 4\mathbf{i} - 8\mathbf{j}$

44. Express the vector as a combination of the standard unit vectors **i** and **j**.
$\mathbf{v} = \overrightarrow{AB}$ where $A = (8, 4)$ and $B = (-10, -5)$

Objective 5: Find the direction angles of vectors

Find the direction angle of the vector.

45.

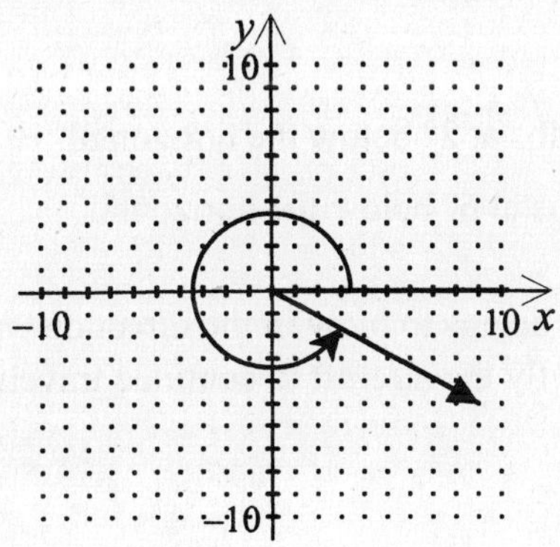

(A) $\theta = 330.9°$ (B) $\theta = 329.9°$ (C) $\theta = 118.1°$ (D) $\theta = 119.1°$

46. $\mathbf{v} = -3\mathbf{i} + 4\mathbf{j}$ (A) $\theta = 53.1°$ (B) $\theta = 233.1°$ (C) $\theta = 36.9°$ (D) $\theta = 126.9°$

47. $\mathbf{v} = \langle 5, 1 \rangle$

48. Find the missing component of the vector with the given direction angle.
$\mathbf{v} = \langle -4\sqrt{2}, y \rangle$, with $\theta = 135°$

Objective 6: Use vectors to model and solve real-life problems

49. A force of 880 pounds is needed to push a stalled car up a hill inclined at an angle of 15° to the horizontal. Find the weight of the car. Ignore friction.

(A) 3284 lb (B) 3081 lb (C) 3387 lb (D) 3400 lb

50. A box is being pulled by means of two ropes. One rope exerts 60 pounds of force at 24° above the horizontal. The other rope exerts 66 pounds of force at 18° below the horizontal. Find the direction and magnitude of the resultant of the two forces.

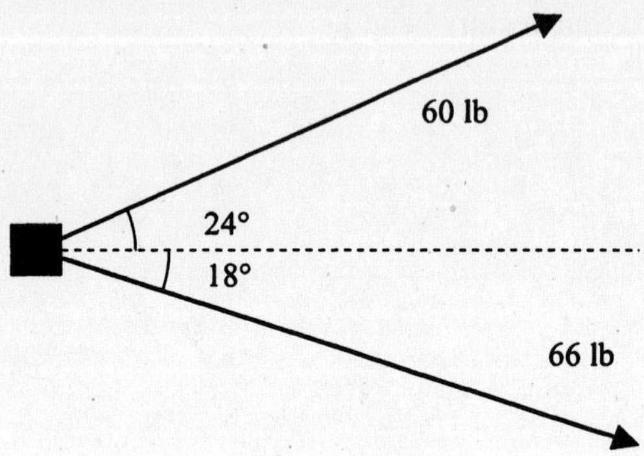

(A) 45.5 lbs at 6° above the horizontal (B) 117.7 lbs at 2° below the horizontal

(C) 117.7 lbs at 2° above the horizontal (D) 45.5 lbs at 6° below the horizontal

51. An airplane is flying due south at 450 miles per hour. A wind begins to blow in the direction N 45° E at 42 miles per hour. Find the bearing the pilot must fly the aircraft to continue traveling due south.

52. A pendulum bob is held stationary, as shown, by a force of 33.3 newtons. Find the tension in the pendulum string.

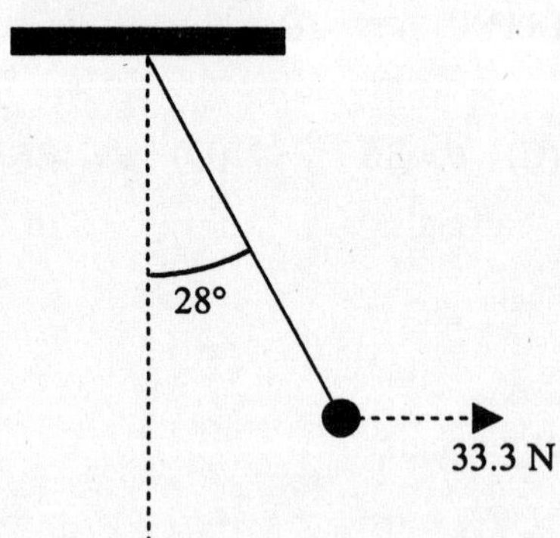

Section 3.4: Vectors and Dot Products

Objective 1: Find the dot product of two vectors and use the Properties of the Dot Product

53. Use the given vectors to find the indicated quantities.

$\mathbf{u}=\langle -4,\ 1\rangle,\ \mathbf{v}=\langle 5,\ -1\rangle,\ \mathbf{w}=\langle 3,\ -4\rangle$

(a) $(\mathbf{u}\cdot\mathbf{v})\mathbf{w}$ (b) $5\mathbf{u}\cdot\mathbf{v}$

(A) (a) $\langle 27,\ -36\rangle$ (b) 45
(B) (a) $\langle -27,\ 36\rangle$ (b) -45
(C) (a) $\langle -63,\ 84\rangle$ (b) -105
(D) (a) $\langle -57,\ 76\rangle$ (b) -95

54. Find the indicated dot product.

$\mathbf{v}=-5\mathbf{i}-8\mathbf{j}$ and $\mathbf{w}=24\mathbf{i}-15\mathbf{j}$, find $\mathbf{v}\cdot\mathbf{w}$

(A) –320 (B) 0 (C) 400 (D) –240

55. Use the properties of the dot product to find $\|\mathbf{u}\|$.

$\mathbf{u}\cdot\mathbf{u}=578$

56. Use the properties of the dot product to evaluate the expression.

$\mathbf{u}=\langle 3,\ 2\rangle,\ \mathbf{v}=\langle 3,\ -4\rangle,\ \mathbf{w}=\langle -3,\ 3\rangle$, find $\mathbf{v}\cdot(\mathbf{w}-\mathbf{u})$

Objective 2: Find the angle between two vectors and determine whether two vectors are orthogonal

57. Find the angle between the two vectors.

$$\mathbf{u}=\cos\left(\frac{5\pi}{6}\right)\mathbf{i}+\sin\left(\frac{5\pi}{6}\right)\mathbf{j}$$

$$\mathbf{v}=\cos\left(\frac{3\pi}{4}\right)\mathbf{i}+\sin\left(\frac{3\pi}{4}\right)\mathbf{j}$$

(A) $\theta=\frac{5\pi}{6}$ (B) $\theta=\frac{3\pi}{4}$ (C) $\theta=\frac{\pi}{6}$ (D) $\theta=\frac{\pi}{12}$

58. Find two vectors in opposite directions that are orthogonal to the vector **u**.

$$\mathbf{u}=\left\langle \frac{7}{25},\ \frac{24}{25}\right\rangle$$

(A) $\left\langle \frac{24}{25},\ -\frac{7}{25}\right\rangle,\ \left\langle \frac{14}{25},\ \frac{48}{25}\right\rangle$
(B) $\left\langle \frac{24}{25},\ -\frac{7}{25}\right\rangle,\ \left\langle \frac{48}{25},\ -\frac{14}{25}\right\rangle$
(C) $\left\langle \frac{24}{25},\ -\frac{7}{25}\right\rangle,\ \left\langle -\frac{48}{25},\ \frac{14}{25}\right\rangle$
(D) $\left\langle \frac{14}{25},\ \frac{48}{25}\right\rangle,\ \left\langle -\frac{24}{25},\ \frac{7}{25}\right\rangle$

59. Find the angle between the two vectors. Express your answer in radians rounded to the nearest tenth.

$\mathbf{u} = -4\mathbf{i} - 8\mathbf{j}$

$\mathbf{v} = -7\mathbf{i} - 5\mathbf{j}$

60. Determine if the vectors are orthogonal, parallel, or neither.

$-8\mathbf{i} - 4\mathbf{j}$ and $8\mathbf{i} - 16\mathbf{j}$

Objective 3: Write a vector as the sum of two vector components

61. Identify the decomposition of **u** into the sum of two orthogonal vectors, one of which is $\text{proj}_{\mathbf{v}}\mathbf{u}$.

$\mathbf{u} = \langle -5, -2 \rangle$, $\mathbf{v} = \langle -8, -5 \rangle$

(A) $\mathbf{u} = \left\langle -\frac{400}{89}, -\frac{250}{89} \right\rangle + \left\langle -\frac{45}{89}, \frac{72}{89} \right\rangle$

(B) $\mathbf{u} = \left\langle -\frac{403}{89}, \frac{67}{89} \right\rangle + \left\langle -\frac{42}{89}, -\frac{245}{89} \right\rangle$

(C) $\mathbf{u} = \left\langle -\frac{45}{89}, -\frac{250}{89} \right\rangle + \left\langle -\frac{400}{89}, \frac{72}{89} \right\rangle$

(D) $\mathbf{u} = \left\langle -\frac{403}{89}, -\frac{245}{89} \right\rangle + \left\langle -\frac{42}{89}, \frac{67}{89} \right\rangle$

62. A car weighing 3400 pounds is parked on the side of a hill with a 10° slope. Assume the only force to overcome is gravity.
(a) Find the force needed to keep the car from rolling down the hill.
(b) Find the force perpendicular to the slope of the hill.

(A) (a) 253 pounds (b) 3348 pounds
(B) (a) 590 pounds (b) 3348 pounds
(C) (a) 253 pounds (b) 3391 pounds
(D) (a) 590 pounds (b) 3391 pounds

63. Find the projection of $\mathbf{v} = -13\mathbf{i} + 13\mathbf{j}$ onto $\mathbf{u} = 2\mathbf{i} + 3\mathbf{j}$ and the vector component of **v** orthogonal to **u**.

64. Kim pushes a lawn spreader across a lawn by applying a force of 89 newtons along the handle, which makes an angle of 55° with the horizontal.
(a) What are the horizontal and vertical components of the force?
(b) The handle is lowered so it makes an angle of 30° with the horizontal. What are the horizontal and vertical components of this force?

Objective 4: Use vectors to find the work done by a force

65. A force of 15 pounds is applied at an angle of 30° above the horizontal to push a cart across the floor. Find the work done if the cart is moved 25 feet.

(A) 188 foot-pounds (B) 325 foot-pounds (C) 125 foot-pounds (D) 375 foot-pounds

66. A horizontal force of 115 pounds is applied to an object as it is pushed up a ramp that is 20 feet long. Find the work done if the ramp is inclined at an angle of 20° above the horizontal.

(A) 787 foot-pounds (B) 2161 foot-pounds (C) 511 foot-pounds (D) 2300 foot-pounds

67. Find the work done in moving a particle from P to Q if the magnitude and direction of the force are given by **v**.

$P=(2,\ 4),\ Q=(6,\ 7),\ \mathbf{v}=\langle 1,\ 7\rangle$

68. A force of 19 pounds is applied to a push lawn mower, at an angle of 50° with the horizontal. Find the work done if the lawn mower is moved 7 feet. Round your answer to the nearest tenth.

Answer Key for Chapter 3 Additional Topics in Trigonometry

Section 3.1: Law of Sines

Objective 1: Use the Law of Sines to solve oblique triangles (AAS or ASA)

[1] (D)

[2] (C)

[3] 35.7

[4] 9.04

Objective 2: Use the Law of Sines to solve oblique triangles (SSA)

[5] (C)

[6] (C)

[7] No solutions exist.

[8] Answers may vary. Sample answer: $a = 17$ or $a > 15.5$

Objective 3: Find the areas of oblique triangles

[9] (D)

[10] (D)

[11] $c = 7$

[12] There are no solutions.

Objective 4: Use the Law of Sines to model and solve real-life problems

[13] (D)

[14] (D)

[15] 2854 m

[16] 134.4°

Section 3.2: Law of Cosines

Objective 1: Use the Law of Cosines to solve oblique triangles (SSS or SAS)

[17] (C)

[18] (A)

[19] 19.9

[20] 66.5°

Objective 2: Use the Law of Cosines to model and solve real-life problems

[21] (D)

[22] (A)

[23] 51.9°

[24] $d^2 = 295^2 + 340^2 - 2(295)(340)\cos 17°$; $d \approx 103.9$ yards

Objective 3: Use Heron's Area Formula to find the area of a triangle

[25] (C)

[26] (B)

[27] 398.2

[28] 25.9 square units

Section 3.3: Vectors in the Plane

Objective 1: Represent vectors as directed line segments

[29] (B)

[30] (D)

[31] $\|\mathbf{u}\| = \sqrt{(-8-(-3))^2 + (5-9)^2} = \sqrt{41}$, slope $= \frac{4}{5}$

$\|\mathbf{v}\| = \sqrt{(-8-(-3))^2 + (-8-(-4))^2} = \sqrt{41}$, slope $= \frac{4}{5}$

[32] $k = -3$

Objective 2: Write the component forms of vectors

[33] (C)

[34] (B)

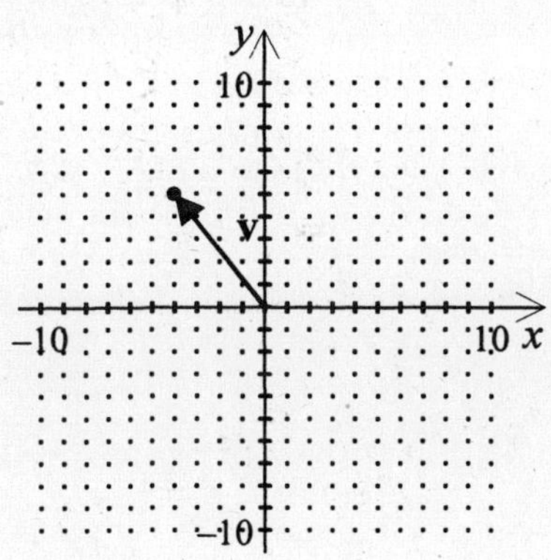

[35] $\mathbf{v} = \langle -4, 5 \rangle$

[36] $y = \pm 7$

Objective 3: Perform basic vector operations and represent them graphically

[37] (B)

[38] (B)

[39] $\langle -12, 4\rangle$

[40] $\left\langle -\frac{6}{7}, -\frac{5}{7}\right\rangle$

Objective 4: Write vectors as linear combinations of unit vectors

[41] (C)

[42] (B)

[43] $\frac{\sqrt{5}}{5}\mathbf{i} - \frac{2\sqrt{5}}{5}\mathbf{j}$

[44] $-18\mathbf{i} - 9\mathbf{j}$

Objective 5: Find the direction angles of vectors

[45] (A)

[46] (D)

[47] $\theta = 11.3°$

[48] $y = 4\sqrt{2}$

Objective 6: Use vectors to model and solve real-life problems

[49] (D)

[50] (C)

[51] S 3.8° W

[52] 71 N

Section 3.4: Vectors and Dot Products

Objective 1: Find the dot product of two vectors and use the Properties of the Dot Product

[53] (C)

[54] (B)

[55] $\|\mathbf{u}\| = 17\sqrt{2}$

[56] –22

Objective 2: Find the angle between two vectors and determine whether two vectors are orthogonal

[57] (D)

[58] (C)

[59] 0.5 radians

[60] Orthogonal

Objective 3: Write a vector as the sum of two vector components

[61] (A)

[62] (B)

[63] $2\mathbf{i} + 3\mathbf{j};\ -15\mathbf{i} + 10\mathbf{j}$

(a) 51 N, 73 N
[64] (b) 77 N, 45 N

Objective 4: Use vectors to find the work done by a force

[65] (B)

[66] (B)

[67] 25

[68] 85.5 foot-pounds

Chapter 4 Complex Numbers

Section 4.1: Complex Numbers

Objective 1: Use the imaginary unit i to write complex numbers

1. Express $\sqrt{-31}$ in the form bi where b is a real number.

(A) $-31i$ (B) $\sqrt{-31}i$ (C) $-\sqrt{31}i$ (D) $\sqrt{31}i$

2. Which is the complex number in standard form?

$5i-12i^2$

(A) $-12+5i$ (B) $17i$ (C) $12-5i$ (D) $12+5i$

3. What real numbers a and b make the equation true?

$(a-1)+(b-9)i = 8-4i$

4. Write the expression $18-\sqrt{-343}$ as a complex number.

Objective 2: Add, subtract, and multiply complex numbers

Perform the indicated operation and write the result in standard form.

5. $(2-i)(-7-9i)$ (A) $-5-11i$ (B) $-5-25i$ (C) $-23+11i$ (D) $-23-11i$

6. $(8+5i)-(-7+9i)$ (A) $-101+37i$ (B) $1-14i$ (C) $1+14i$ (D) $15-4i$

7. $(-2-4i)^2$

8. $\sqrt{-24}+\sqrt{-54}$

Objective 3: Use complex conjugates to divide complex numbers

Divide and write the result in standard form.

9. $\dfrac{-2+6i}{9i}$ (A) $\dfrac{-54-2i}{9}$ (B) $\dfrac{6+2i}{81}$ (C) $\dfrac{6+2i}{9}$ (D) $\dfrac{-54-2i}{81}$

Divide and write the result in standard form.

10. $\dfrac{-1-8i}{3-6i}$ (A) $\dfrac{3-2i}{6}$ (B) $\dfrac{-51+2i}{6}$ (C) $\dfrac{-51+2i}{3}$ (D) $\dfrac{3-2i}{3}$

11. $\dfrac{8+3i}{1+5i}$

12. $\dfrac{-3i}{(2+i)^2}$

Objective 4: Use the Quadratic Formula to find complex solutions of quadratic equations

Use the Quadratic Formula to solve.

13. $-\dfrac{4}{5}x^2-\dfrac{7}{2}x-10=0$

(A) $-\dfrac{35}{16}\pm\dfrac{5}{16}\sqrt{79}\,i$ (B) $\dfrac{2}{3}\pm\dfrac{17}{3}\sqrt{\dfrac{4}{5}}\,i$ (C) $\dfrac{35}{16}\pm\dfrac{5}{16}\sqrt{79}\,i$ (D) $-\dfrac{2}{3}\pm\dfrac{17}{3}\sqrt{\dfrac{4}{5}}\,i$

14. $2x^2+45=0$

(A) $x=\dfrac{\pm 3\sqrt{5}i}{2}$ (B) $x=\pm 3\sqrt{5}i$ (C) $x=\dfrac{\pm 3\sqrt{10}i}{2}$ (D) $x=\dfrac{\pm\sqrt{10}i}{15}$

15. $x^2-8x+20=0$

16. $-x+6+4x^2=0$

Section 4.2: Complex Solutions of Equations

Objective 1: Determine the numbers of solutions of polynomial equations

Determine the maximum number of zeros of the polynomial function.

17. $f(x)=\left(2x^4-8x^2-3\right)\left(x^3+9x+1\right)$ (A) 10 (B) 7 (C) 4 (D) 9

Determine the maximum number of zeros of the polynomial function.

18. $f(x) = -3x^3 + 5x^2 + 7x + 10$ (A) 5 (B) 3 (C) –3 (D) 2

19. $f(x) = x^2(x+4)^4$

20. $f(x) = 2x^4(5x^3 - 3x^2 + 2x + 7)$

Objective 2: Find solutions of polynomial equations

21. Find all the real zeros of the function.
$f(x) = 2x^4 - 5x^3 - 29x^2 + 8x + 60$

(A) $-\frac{3}{2}, \frac{1}{5}, 5, -2$ (B) $-\frac{3}{2}, 2, -5$ (C) $\frac{3}{2}, -2, 5$ (D) $\frac{3}{2}, -\frac{1}{5}, 5, -2$

22. Find all the zeros of the function.
$f(x) = x^4 - 6x^3 + 6x^2 + 24x - 40$

(A) $\pm 2,\ 3 \pm 2i$ (B) $\pm 3,\ 2 \pm 2i$ (C) $\pm 2,\ 3 \pm i$ (D) none of these

23. Find all the real zeros of the function.
$f(x) = 45x^3 - 36x^2 - 5x + 4$

24. Find all the zeros of the function.
$f(x) = x^4 + 2x^3 + x^2 - 2x - 2$

Objective 3: Find the zeros of polynomial functions

25. Identify which of the following is a third-degree polynomial function with real coefficients and with zeros at -2 and $2+i$.

(A) $P(x) = x^3 - 6x^2 + 13x - 10$ (B) $P(x) = x^3 - 2x^2 + 13x + 10$

(C) $P(x) = x^3 + 2x^2 - 3x - 10$ (D) $P(x) = x^3 - 2x^2 - 3x + 10$

26. A polynomial of degree 5 whose coefficients are real numbers has the zeros 2, $8i$, and $8+i$. Identify the remaining zeros.

(A) $-8i,\ 8+i$ (B) $-2,\ -8i,\ 8-i$ (C) $-8i,\ 8-i$ (D) $8i,\ -8-i$

27. Find a quadratic function with solutions $2+2i$ and $2-2i$ and a leading coefficient of 1.

28. Given that one zero of $P(x)=x^3-4x^2+9x-10$ is $1+2i$, find all zeros of $P(x)$.

Section 4.3: Trigonometric Form of a Complex Number

Objective 1: Plot complex numbers in the complex plane

29. Which number is farthest from the origin in the complex plane?

(A) $6+7i$ (B) $4+2i$ (C) $-7-7i$ (D) $-2-8i$

30. Match the complex number with its graph.
$4-8i$

(A)

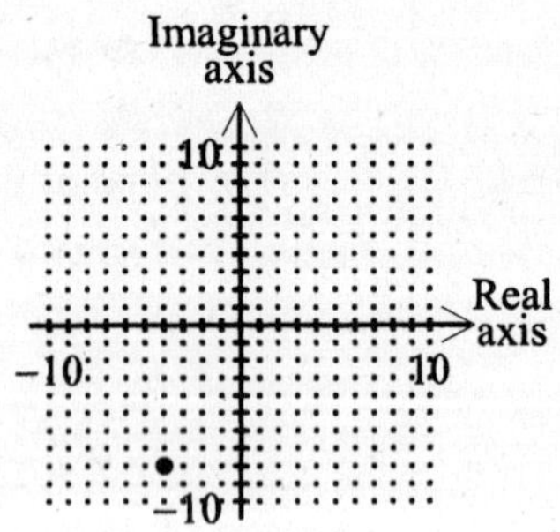

(B)

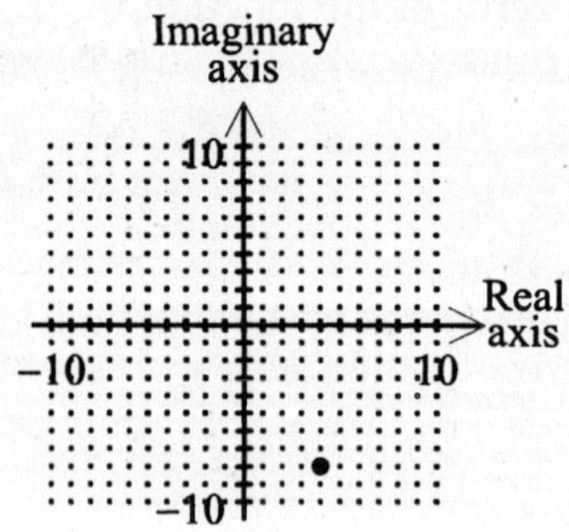

(C)

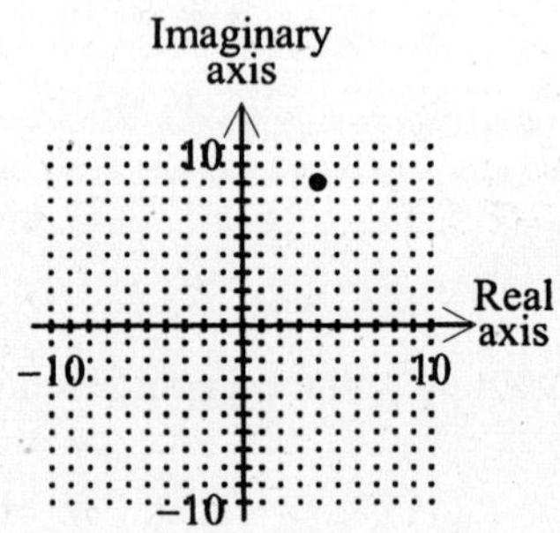

(D)

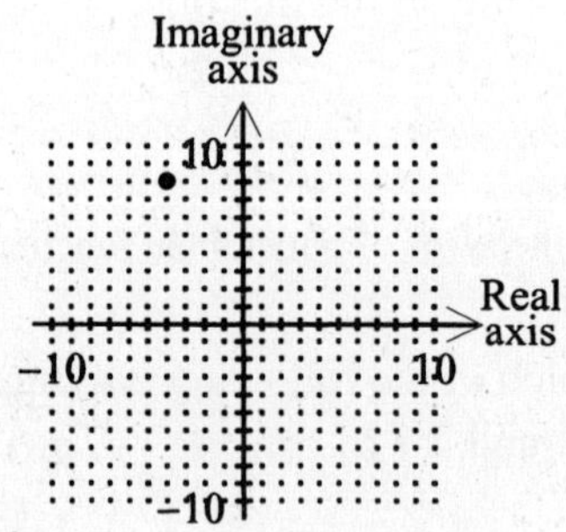

31. Plot the complex number.
$-7-i$

32. Write the complex number in standard form.

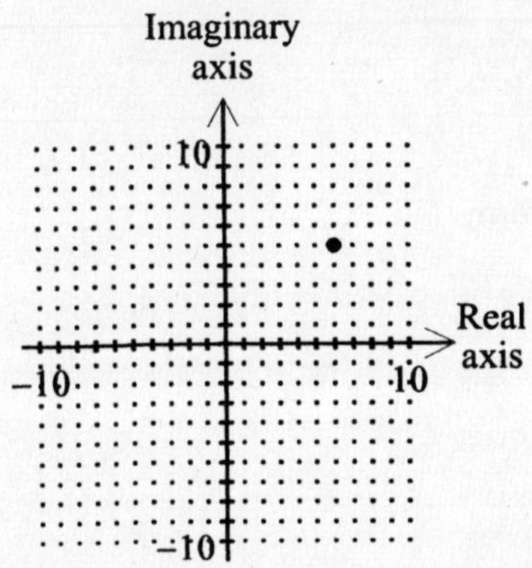

Objective 2: Write the trigonometric forms of complex numbers

33. Identify the standard form of the complex number.
$15(\cos 269° + i\sin 269°)$

(A) $-0.26-15.00i$ (B) $-1.00-0.02i$ (C) $-0.26-0.02i$ (D) $-1.00-15.00i$

34. Identify the trigonometric form of the complex number.
$8-i$

(A) $8.06(\cos 352.9° - i\sin 352.9°)$ (B) $8.06(\cos 352.9° + i\sin 352.9°)$

(C) $7(\cos 172.9° + i\sin 172.9°)$ (D) $65(\cos 172.9° - i\sin 172.9°)$

35. Write the complex number in trigonometric form.

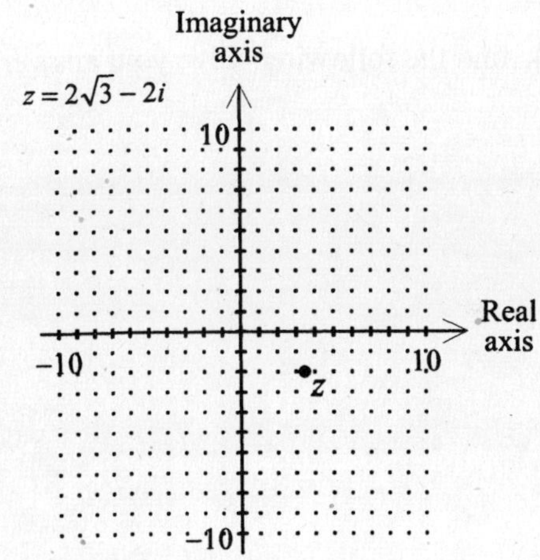

36. Find the exact modulus of the trigonometric form of the complex number z. Then find its argument to the nearest tenth of a degree.
$z = -5 - 1i$

Objective 3: Multiply and divide complex numbers written in trigonometric form

37. Find zw.
$z = 9\left(\cos\frac{4\pi}{9} + i\sin\frac{4\pi}{9}\right),\ w = 3\left(\cos\frac{\pi}{9} + i\sin\frac{\pi}{9}\right)$

(A) $27\left(\cos\frac{80\pi}{9} + i\sin\frac{80\pi}{9}\right)$
(B) $27\left(\cos\frac{5\pi}{9} + i\sin\frac{5\pi}{9}\right)$
(C) $12\left(\cos\frac{\pi}{3} + i\sin\frac{\pi}{3}\right)$
(D) none of these

38. Find the value of the expression.
$$\frac{(\cos 105° + i\sin 105°)}{(\cos 125° + i\sin 125°)}$$

(A) $0.940 - 0.342i$ (B) $-0.94 - 0.342i$ (C) $0.342 - 0.94i$ (D) $-0.342 + 0.94i$

39. Perform the multiplication. Give your answer in both trigonometric form and standard form. Do not estimate.
$5\left(\cos\frac{\pi}{3} + i\sin\frac{\pi}{3}\right)\cdot 2\left(\cos\frac{4\pi}{3} + i\sin\frac{4\pi}{3}\right)$

40. If $z = 6(\cos 60° + i\sin 60°)$ and $w = 3(\cos 15° + i\sin 15°)$, find the following. Give your answer in trigonometric form.
(a) zw
(b) $\frac{z}{w}$

Section 4.4: DeMoivre's Theorem

Objective 1: Use DeMoivre's Theorem to find powers of complex numbers

Use DeMoivre's theorem to find the indicated power of the complex number.

41. $\left(\sqrt{2}+\sqrt{2}i\right)^2$ (A) 4 (B) $-4i$ (C) $4i$ (D) -4

42. $(9+3i)^3$ (A) $702+486i$ (B) $23.4+16.2i$ (C) $16.2+23.4i$ (D) $486+702i$

43. $(-2-2i)^5$

44. Use DeMoivre's theorem to find the indicated power of the complex number. Express the result in standard form. Do not approximate.
$\left[4(\cos 15° + i \sin 15°)\right]^2$

Objective 2: Find nth roots of complex numbers

Find the indicated roots of the complex number.

45. Cube roots of $-1-i$

(A) $8(\cos 75° + i\sin 75°)$, $8(\cos 195° + i\sin 195°)$, $8(\cos 315° + i\sin 315°)$

(B) $\sqrt{2}(\cos 75° + i\sin 75°)$, $\sqrt{2}(\cos 195° + i\sin 195°)$, $\sqrt{2}(\cos 315° + i\sin 315°)$

(C) $\sqrt[6]{2}(\cos 75° + i\sin 75°)$, $\sqrt[6]{2}(\cos 195° + i\sin 195°)$, $\sqrt[6]{2}(\cos 315° + i\sin 315°)$

(D) none of these

46. Fourth roots of $625(\cos 65° + i\sin 65°)$

(A) $43.7233+150.0078i$, $150.0078-43.7233i$, $-43.7233-150.0078i$, $-150.0078+43.7233i$

(B) $1.3991+4.8002i$, $4.8002-1.3991i$, $-1.3991-4.8002i$, $-4.8002+1.3991i$

(C) $4.8002+1.3991i$, $-1.3991+4.8002i$, $-4.8002-1.3991i$, $1.3991-4.8002i$

(D) $150.0078+43.7233i$, $-43.7233+150.0078i$, $-150.0078-43.7233i$, $43.7233-150.0078i$

47. Find all the solutions of the equation.

$x^5 + 243 = 0$

48. Find the indicated roots of the complex number. Express the answers in trigonometric form, with θ in degrees.

Cube roots of $1 + i$

Answer Key for Chapter 4 Complex Numbers

Section 4.1: Complex Numbers

Objective 1: Use the imaginary unit i to write complex numbers

[1] (D)

[2] (D)

[3] $a = 9,\ b = 5$

[4] $18 - 7\sqrt{7}\,i$

Objective 2: Add, subtract, and multiply complex numbers

[5] (D)

[6] (D)

[7] $-12 + 16i$

[8] $5\sqrt{6}i$

Objective 3: Use complex conjugates to divide complex numbers

[9] (C)

[10] (D)

[11] $\frac{23}{26} - \frac{37}{26}i$

[12] $-\frac{12}{25} - \frac{9}{25}i$

Objective 4: Use the Quadratic Formula to find complex solutions of quadratic equations

[13] (A)

[14] (C)

[15] $4 \pm 2i$

[16] $\dfrac{1 \pm \sqrt{95}i}{8}$

Section 4.2: Complex Solutions of Equations

Objective 1: Determine the numbers of solutions of polynomial equations

[17] (B)

[18] (B)

[19] 6

[20] 7

Objective 2: Find solutions of polynomial equations

[21] (C)

[22] (C)

[23] $\frac{4}{5}, \frac{1}{3}, -\frac{1}{3}$

[24] $\pm 1, -1 \pm i$

Objective 3: Find the zeros of polynomial functions

[25] (D)

[26] (C)

[27] $f(x) = x^2 - 4x + 8$

[28] $1 \pm 2i,\ 2$

Section 4.3: Trigonometric Form of a Complex Number

Objective 1: Plot complex numbers in the complex plane

[29] (C)

[30] (B)

[31]

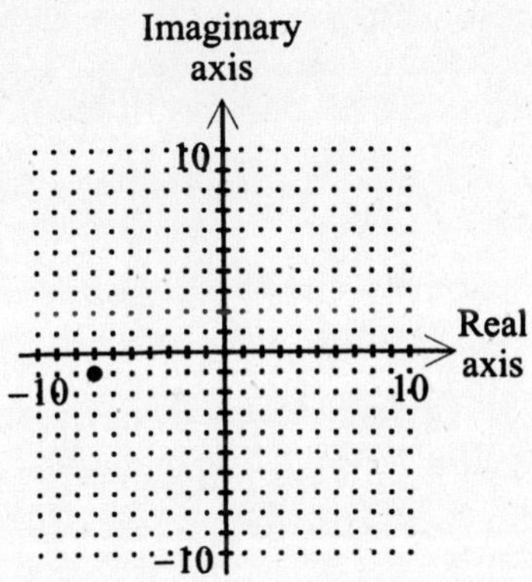

[32] $6 + 5i$

Objective 2: Write the trigonometric forms of complex numbers

[33] (A)

[34] (B)

[35] $4\left(\cos \frac{11\pi}{6} + i \sin \frac{11\pi}{6}\right)$

[36] Modulus $= \sqrt{26}$; argument $= 191.3°$

Objective 3: Multiply and divide complex numbers written in trigonometric form

[37] (B)

[38] (A)

[39] $10\left(\cos\frac{5\pi}{3}+i\sin\frac{5\pi}{3}\right)=5-5\sqrt{3}i$

[40] (a) $zw=18\left(\cos 75^\circ+i\sin 75^\circ\right)$

(b) $\frac{z}{w}=2\left(\cos 45^\circ+i\sin 45^\circ\right)$

Section 4.4: DeMoivre's Theorem

Objective 1: Use DeMoivre's Theorem to find powers of complex numbers

[41] (C)

[42] (D)

[43] $128+128i$

[44] $8\sqrt{3}+8i$

Objective 2: Find nth roots of complex numbers

[45] (C)

[46] (C)

[47] $2.4271+1.7634i,\ -0.9271+2.8532i,\ -3,\ -0.9271-2.8532i,\ 2.4271-1.7634i$

[48] $\sqrt[6]{2}(\cos 15^\circ+i\sin 15^\circ),\ \sqrt[6]{2}(\cos 135^\circ+i\sin 135^\circ),\ \sqrt[6]{2}(\cos 255^\circ+i\sin 255^\circ)$

Chapter 5 Exponential and Logarithmic Functions

Section 5.1: Exponential Functions and Their Graphs

Objective 1: Recognize and evaluate exponential functions with base a

Evaluate the expression.

1. $3500\left(2^{-1.9}\right)$ (A) –13,300   (B) 12,635   (C) 1061.252   (D) 937.802

2. $5^{-2\pi}$ (A) 0.000041   (B) –31.416   (C) –1.283   (D) 24,646.659

Evaluate the expression. Round the result to three decimal places.

3. $4^{1.8}$

4. $1.3^{-3/7}$

Objective 2: Graph exponential functions

Identify the graph of the function.

5. $f(x) = 2^{-2-2x}$

(A)

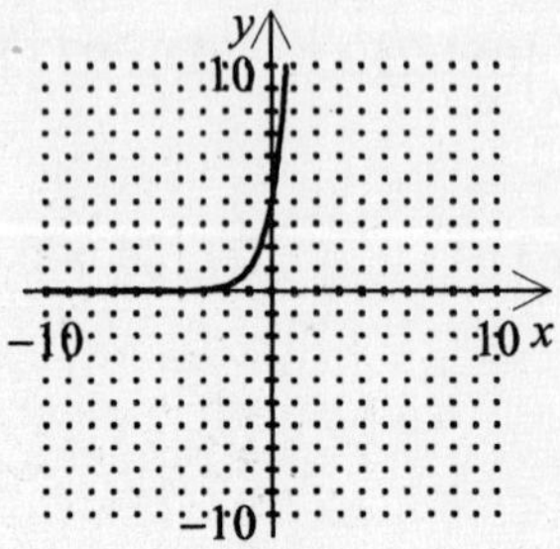

(B)

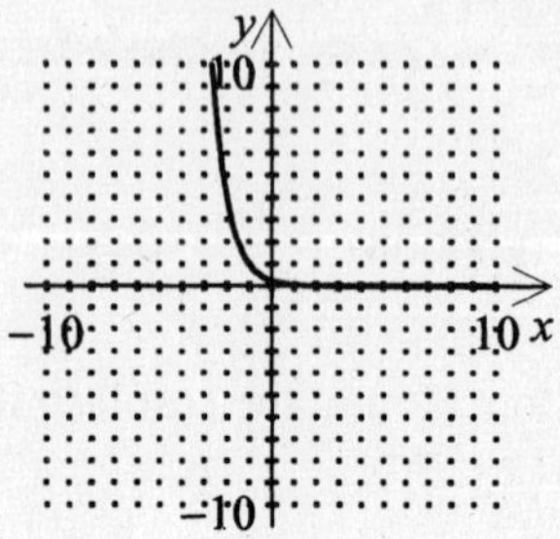

(C)

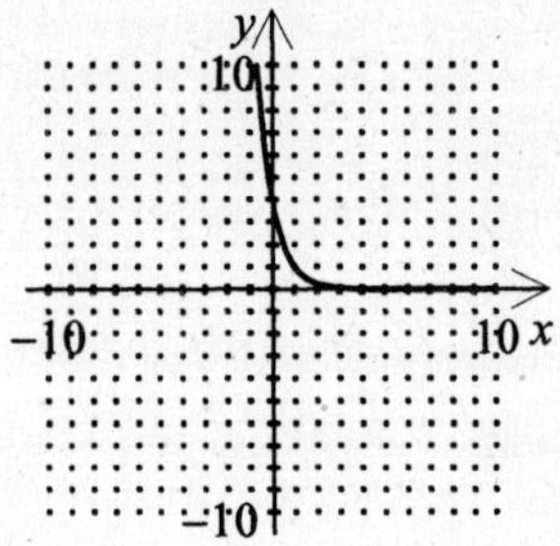

(D)

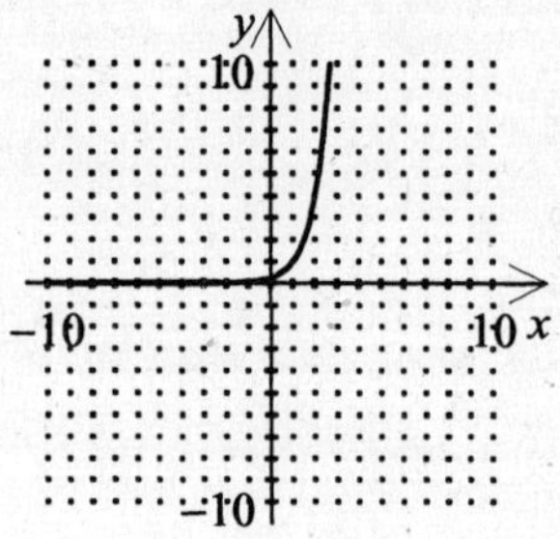

Identify the graph of the function.

6. $f(x) = \left(\frac{1}{10}\right)^x$

(A)

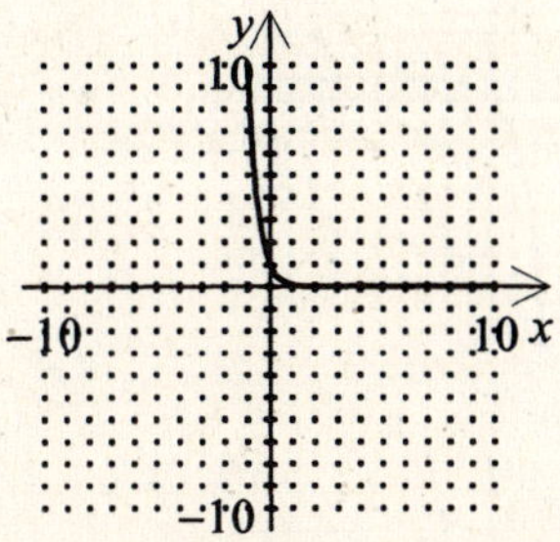

(B)

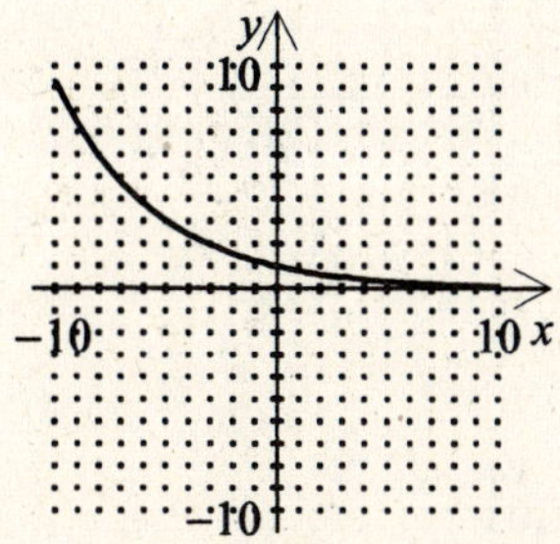

(C)

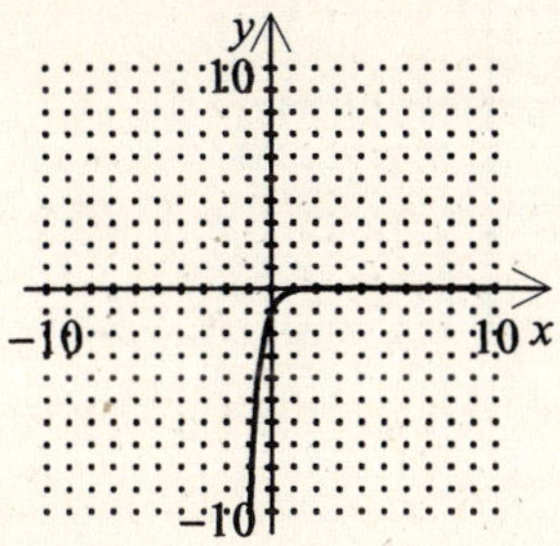

(D)

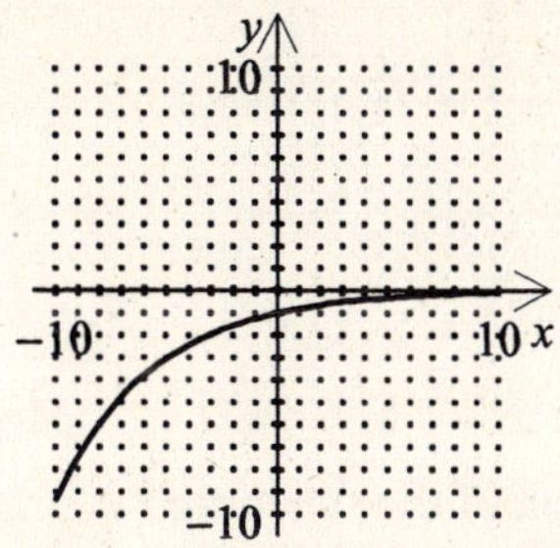

Sketch the graph of the function.

7. $f(x) = \left(\frac{5}{3}\right)^x$

8. $f(x) = \left(\frac{1}{3}\right)^{x-2} - 4$

Objective 3: Recognize and evaluate exponential functions with base e

9. Evaluate the expression. (A) 7.389 (B) 5.437 (C) 0.135 (D) 8.623

e^2

10. Identify the graph of the function.

$f(x) = 1 - e^x$

(A)

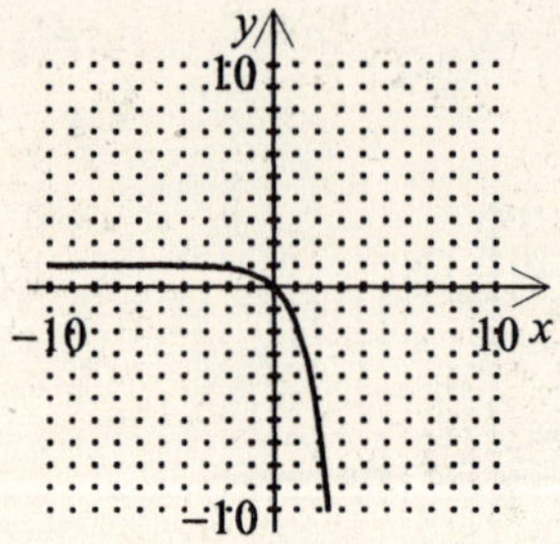

(B)

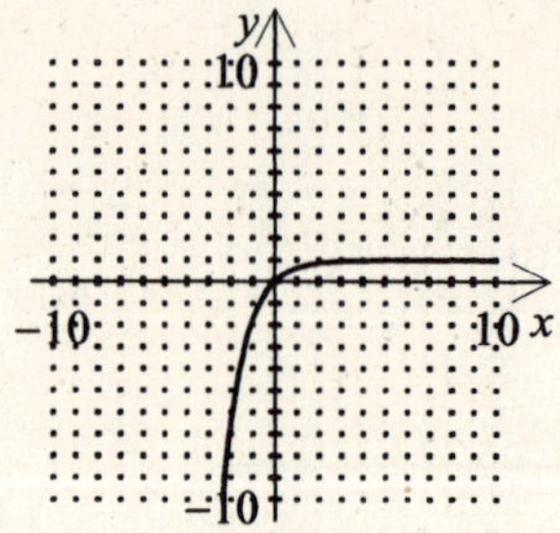

(C)

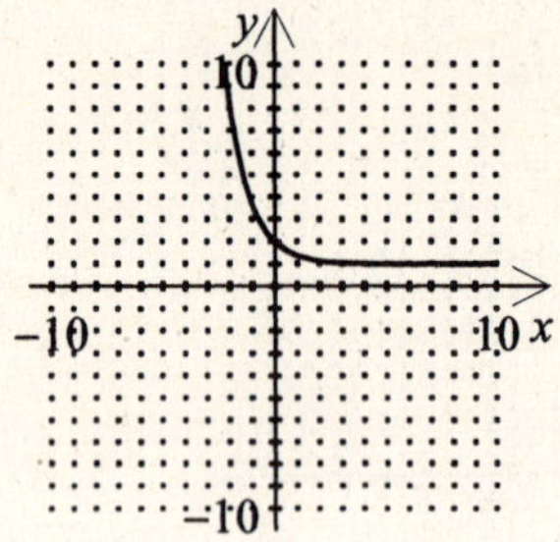

(D)

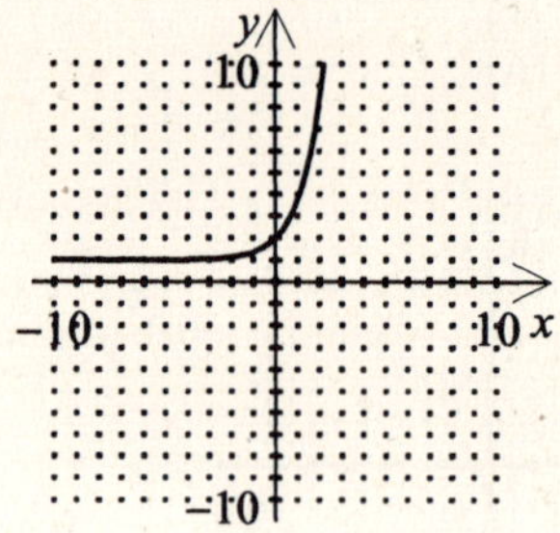

11. Sketch the graph of the function.

$f(x) = 2 - e^x$

12. Evaluate the expression. Round the answer to three decimal places.

$\frac{3}{4}e^{0.5}$

Objective 4: Use exponential functions to model and solve real-life applications

13. If \$4000 is invested in a long-term trust fund with an interest rate of 12% compounded continuously, what is the amount of money in the account after 25 years?

(A) \$68,000.26 (B) \$29,556.22 (C) \$90,585.52 (D) \$80,342.15

14. In 1997, the population of a country was estimated at 4 million. For any subsequent year the population $P(t)$ in millions is

$$P(t) = \frac{240}{5 + 54.99e^{-0.0208t}}$$

where t is the number of years since 1997. Use a graphing calculator to estimate the population in 2016.

(A) 5,762,000 (B) 5,605,000 (C) 5,709,000 (D) 5,815,000

15. The half-life of a radioactive substance is 13.7 years. The quantity of the substance present after t years is

$Q = A_0 e^{-t\ln 2/13.7}$

where A_0 is the original amount in the sample. How much of a 450-gram sample of the substance will remain after 55.0 years? Round the answer to the nearest tenth.

16. The average atmospheric pressure P in pounds per square inch is

$P = 14.7e^{-0.21x}$

where x is the altitude in miles above sea level. Find the average atmospheric pressure for an altitude of 6.9 miles. Round the answer to the nearest tenth.

Section 5.2: Logarithmic Functions and Their Graphs

Objective 1: Recognize and evaluate logarithmic functions with base a

17. Evaluate the expression without using a calculator.

$\log_{81} 3$

(A) –4 (B) $\frac{1}{4}$ (C) 4 (D) $-\frac{1}{4}$

18. Identify the logarithmic equation written in exponential form.

$\log_{125} 25 = \frac{2}{3}$

(A) $\left(\frac{2}{3}\right)^{125} = 25$ (B) $25^{2/3} = 125$ (C) $\left(\frac{2}{3}\right)^{25} = 125$ (D) $125^{2/3} = 25$

19. Write the exponential equation in logarithmic form.

$2^5 = 32$

20. Evaluate the expression without using a calculator.

$\log_{10} \frac{1}{100{,}000}$

Objective 2: Graph logarithmic functions

21. Identify the graph of the logarithmic function.
$f(x) = \log_2(x-5)$

(A)

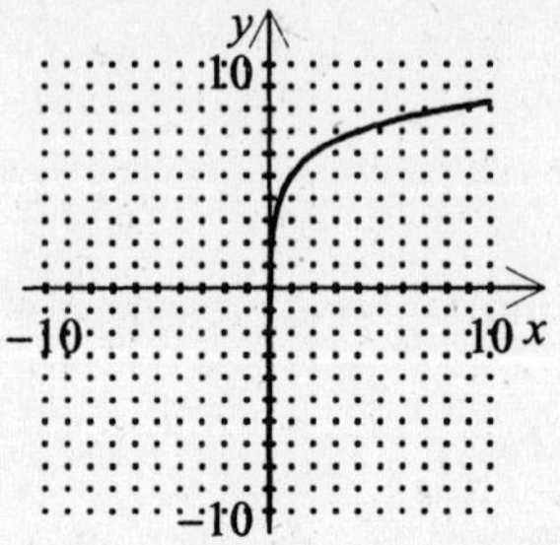

(B)

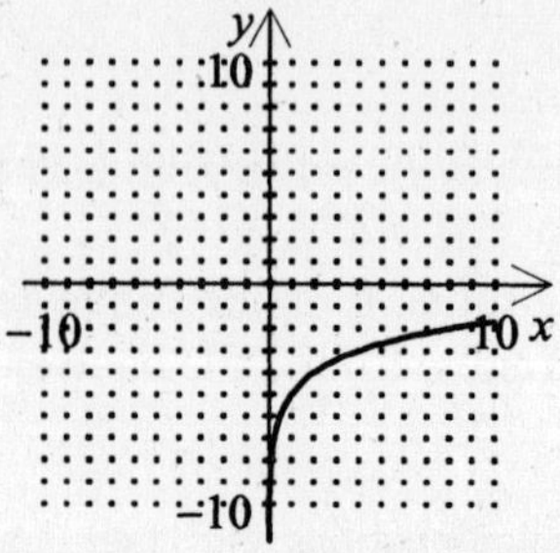

(C)

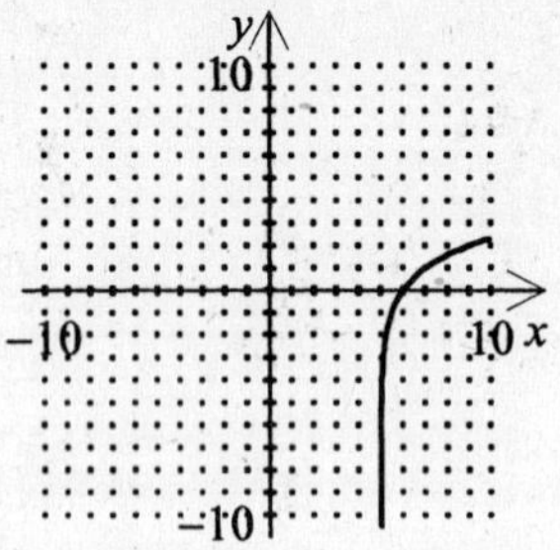

(D)

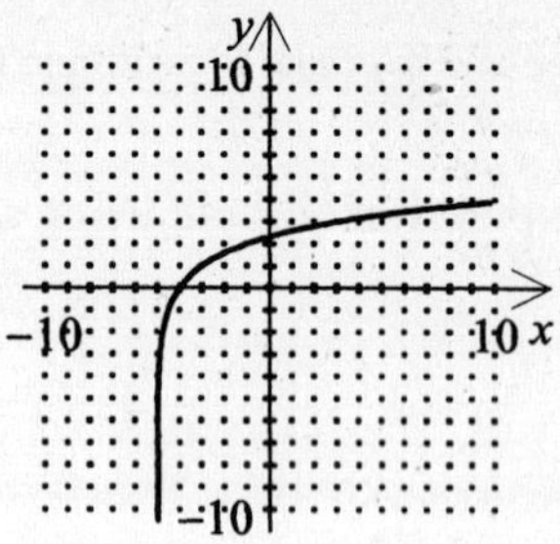

22. Identify the equation represented by the graph.

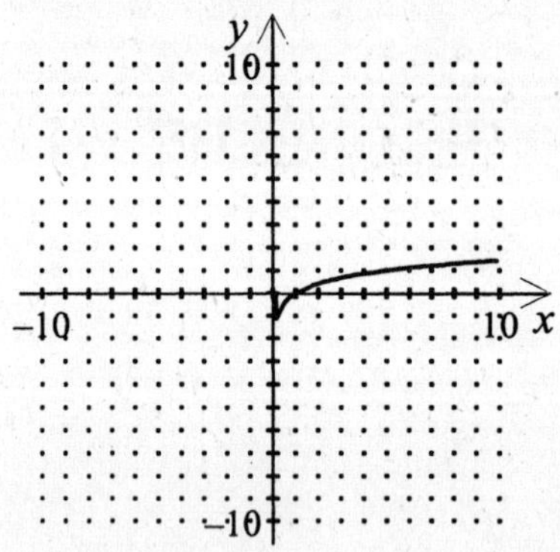

(A) $f(x) = \log_5 x$ (B) $f(x) = \log_x 5$ (C) $f(x) = \log_x \frac{1}{5}$ (D) $f(x) = \log_5 5$

Sketch the graph of the function.

23. $f(x) = \log_4 x$

Sketch the graph of the function.

24. $y = -\log_5 x$

Objective 3: Recognize and evaluate natural logarithmic functions

25. Identify the graph of the logarithmic function.
$f(x) = 3 + \ln(x - 3)$

(A)

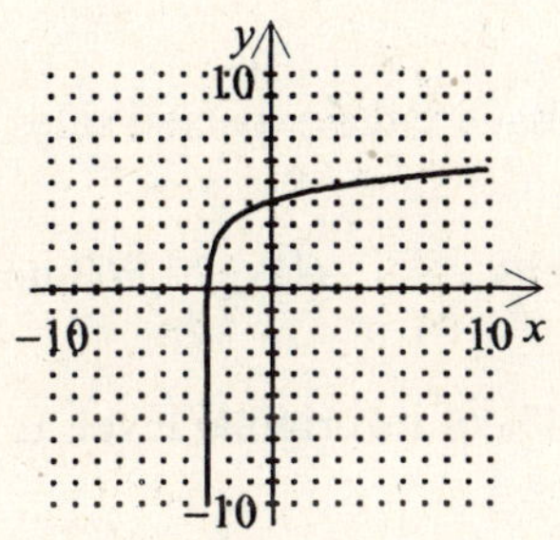

(B)

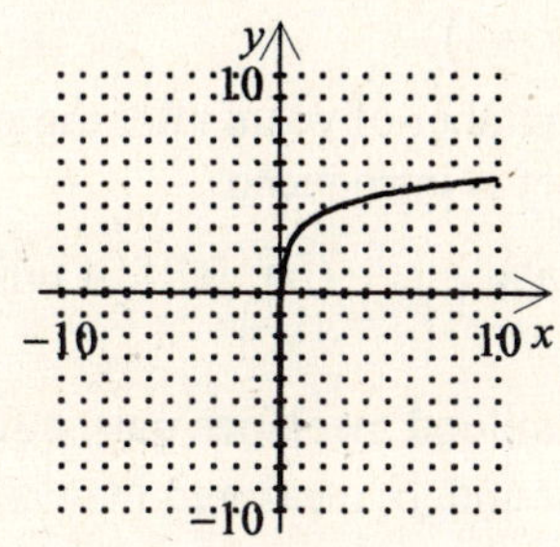

(C)

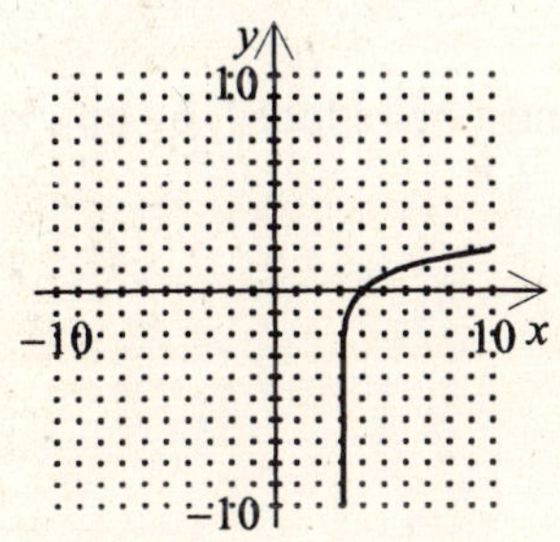

(D)

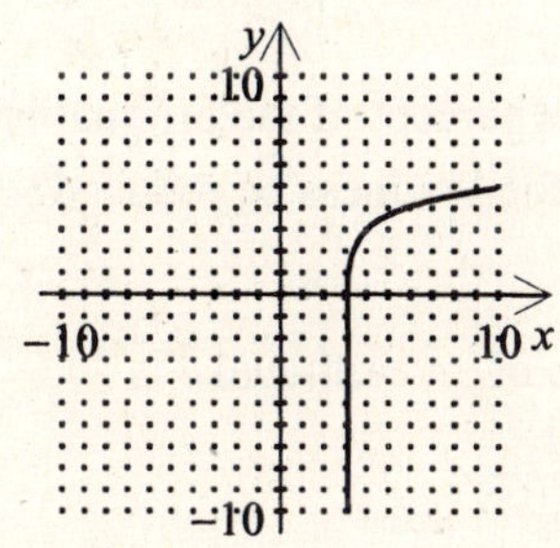

Use a calculator to evaluate the logarithm. Round to three decimal places.

26. ln 39 (A) 3.714 (B) 3.664 (C) 1.591 (D) 6.245

27. $\ln\left(5 + \sqrt{3}\right)$

28. Sketch the graph of the function.
$f(x) = \ln(5 - x)$

Objective 4: Use logarithmic functions to model and solve real-life applications

29. The time required to grow a certain bacteria in a culture beginning with 10 bacteria is

$$t = \frac{\ln B - \ln 10}{0.491}$$

where B is the number of bacteria and t is the time in hours. How much time is required to grow a culture of 9700 bacteria?

(A) 9.7 hours (B) 6.9 hours (C) 14.0 hours (D) 5.8 hours

30. The number of units sold for a certain product is

$$N = 8100 \ln(6t + 4)$$

where t is the number of years after the product is introduced. What are the expected sales 5 years after the product is introduced?

(A) 28,564 units (B) 27,554 units (C) 72,570 units (D) 27,563 units

31. A company with loud machines must cut its sound intensity to 24% of its original level. If the loudness of a sound β measured in decibels is

$$\beta = 10 \log_{10} \frac{100}{I_0}$$

where I_0 is the percent of the original level to which the sound must be reduced, by how many decibels must the loudness be reduced?

32. The magnitude of an earthquake is

$$M = \frac{2}{3} \log_{10} \frac{E}{10^{11.8}}$$

where E is the energy released. If an earthquake released three times as much energy as an earlier quake that released $10^{21.8}$ ergs of energy, find the magnitude of the newer quake to the nearest tenth.

Section 5.3: Properties of Logarithms

Objective 1: Use the change-of-base formula to rewrite and evaluate logarithmic expressions

33. Evaluate the logarithm using the change-of-base formula.

$\log_7 847$

(A) 3.465 (B) 47.192 (C) 0.963 (D) 2.235

34. Which is the logarithm rewritten as a ratio of natural logarithms?

$\log_x \frac{7}{2}$

(A) $\frac{\ln\frac{7}{2}}{\ln x}$ (B) $\ln\left(\frac{\frac{7}{2}}{x}\right)$ (C) $\ln x - \ln\frac{7}{2}$ (D) $\frac{\ln x}{\ln\frac{7}{2}}$

35. Evaluate the logarithm using the change-of-base formula. Find the value to three decimal places.

$\log_6 25$

36. Use the change-of-base formula to rewrite the logarithm as a ratio of logarithms. Then use a graphing utility to sketch the graph.

$f(x) = \log_{5/3} x$

Objective 2: Use properties of logarithms to evaluate or rewrite logarithmic expressions

Find the value of the expression without using a calculator.

37. $6\ln e^3$ (A) $18e$ (B) 18 (C) $\frac{1}{2}$ (D) 2

38. $4\log_3 \sqrt[3]{3}$ (A) 9 (B) $\frac{4}{3}$ (C) 12 (D) $\frac{5}{3}$

39. Indicate whether the following statement is true or false. If it is false, correct the statement.

$\log_{10}(6 \cdot 7) = \log_{10} 6 - \log_{10} 7$

40. Find the value of the expression without using a calculator.

$\log_5 25 + \log_5 625 - \log_5 125$

Objective 3: Use properties of logarithms to expand or condense logarithmic expressions

41. Use the properties of logarithms to expand the expression. (Assume all variables are positive.)

$\log_9 \dfrac{x^3}{\sqrt[6]{y}}$

(A) $\log_9 x^3 + 6\log_9 y$ (B) $\log_9 x^3 - 6\log_9 y$

(C) $3\log_9 x + \dfrac{1}{6}\log_9 y$ (D) $3\log_9 x - \dfrac{1}{6}\log_9 y$

Condense the expression to the logarithm of a single quantity.

42. $3\log_{10} x - 2\log_{10}(x+6)$

(A) $\log_{10} x^3(x+6)^2$ (B) $6\log_{10} \dfrac{x}{x+6}$ (C) $\log_{10} \dfrac{x^3}{(x+6)^2}$ (D) None of these

43. $\left[6\log_6(x+6) + 2\log_6(x+4)\right] - \dfrac{1}{2}\log_6 x$

44. Use the properties of logarithms to expand the expression. (Assume all variables are positive.)

$\log_a \dfrac{4xy^4}{z^3}$

Objective 4: Use logarithmic functions to model and solve real-life applications

45. The amount of time t in years required for a certain radioactive material to decompose is

$t = \dfrac{\ln R - \ln A}{k}$

where R is the mass of the substance remaining after decomposition, A is the original mass of the substance, and k is a constant related to the particular material. Find the time required for 96 grams of a radioactive substance to decompose so that only 16 grams remain if $k = -0.105$.

(A) 170.1 years (B) 43.3 years (C) 0.1 year (D) 17.0 years

46. The magnitude of an earthquake as measured by the Richter Scale is

$R = 0.67\log_{10}(0.37E) + 1.46$

where R is the magnitude of the earthquake and E is the energy in kilowatt-hours released by the earthquake. Which magnitude corresponds to a release of 2.90×10^6 kwh of energy?

(A) 5.2 (B) 6.1 (C) 5.5 (D) None of these

47. The power gain in decibels of an electronic device is determined by the logarithmic expression $10(\log_{10} P_o - \log_{10} P_i)$ where P_o is the output power in watts and P_i is the input power in watts.
(a) Condense the expression using the properties of logarithms.
(b) Determine the power gain of an amplifier with an output of 22.5 W and an input of 0.622 W.

48. The energy E in kilocalories per mole required to transport a substance from the outside to the inside of a living cell is given by
$E = 1.4(\log C_2 - \log C_1)$
where C_1 is the concentration of the substance outside the cell and C_2 is the concentration inside the cell.
(a) Condense the expression using the properties of logarithms.
(b) The concentration of a particular substance outside is four times the concentration inside the cell. How much energy is required to transport the substance from inside to outside the cell? A negative value for E means that energy is needed to prevent the transport of the substance from the region of higher concentration to the region of lower concentration.

Section 5.4: Exponential and Logarithmic Equations

Objective 1: Solve simple exponential and logarithmic equations

Solve for x.

49. $\dfrac{1}{125} = 25^{5x+4}$ (A) $-\dfrac{1}{2}$ (B) $-\dfrac{7}{10}$ (C) $-\dfrac{7}{5}$ (D) $-\dfrac{11}{10}$

50. $\left(\dfrac{1}{2}\right)^x = 128$ (A) $\dfrac{1}{7}$ (B) $-\dfrac{1}{7}$ (C) -7 (D) 7

51. $e^x = \dfrac{1}{4}$

52. $\log_{10} x = 3$

Objective 2: Solve more complicated exponential equations

Solve the exponential equation algebraically.

53. $5e^{0.05x} + 36 = 46$ (A) 40.000 (B) 6.021 (C) 13.863 (D) -0.231

Solve the exponential equation algebraically.

54. $\dfrac{200}{1+e^{-x}}=100$ (A) -0.405 (B) 0.000 (C) -0.082 (D) -1.099

Solve the exponential equation algebraically. Approximate the result(s) to three decimal places.

55. $3^{2x}-11\left(3^{x}\right)+28=0$

56. $85=5e^{0.01t}$

Objective 3: Solve more complicated logarithmic equations

Find the value of x.

57. $4\ln(9x)=23$ (A) 0.639 (B) 0.194 (C) 51.750 (D) 34.910

58. $\log_5(5x+9)=3$ (A) $\dfrac{134}{5}$ (B) $\dfrac{234}{5}$ (C) $\dfrac{116}{5}$ (D) $\dfrac{6}{5}$

59. $\log_2 x+\log_2(x-2)=3$

60. Find the value of x. Round to three decimal places.
$\ln\sqrt{2x-3}=0.9$

Objective 4: Use exponential and logarithmic equations to model and solve real-life applications

61. The number of bacteria present in a culture is
$B=1000e^{0.167t}$
where t is the time in minutes. Find the time required, to the nearest half minute, to have 7400 bacteria present.

(A) 13.0 min (B) 12.0 min (C) 12.5 min (D) 11.5 min

62. An automobile manufacturer is introducing a new fuel-efficient model and estimates the demand for the car as

$N = 54{,}000 \ln(3t + 7)$

where N is the estimated number of cars to be sold and t is the number of years after the car is introduced. When will the demand be 190,000 cars?

(A) 8.9 years (B) 7.2 years (C) 13.6 years (D) 8.5 years

63. The amount of power generated by a satellite's power supply is

$P = 50e^{-t/300}$

where P is the power in watts and t is the time in days. For how many days will 38 watts of power be available? Round to the nearest whole day.

64. The mathematical model for learning an assembly-line procedure needed for assembling one component of a manufactured item is

$P = \dfrac{0.88}{1 + e^{-0.3n}}$

where P is the proportion of correctly assembled components after n practice sessions. How many practice sessions are required to have at least 65% of the components correctly assembled within the given time period? Round any non-integer to the next highest whole number.

Section 5.5: Exponential and Logarithmic Models

Objective 1: Recognize the five most common types of models involving exponential and logarithmic functions

65. Identify the model shown in the graph.

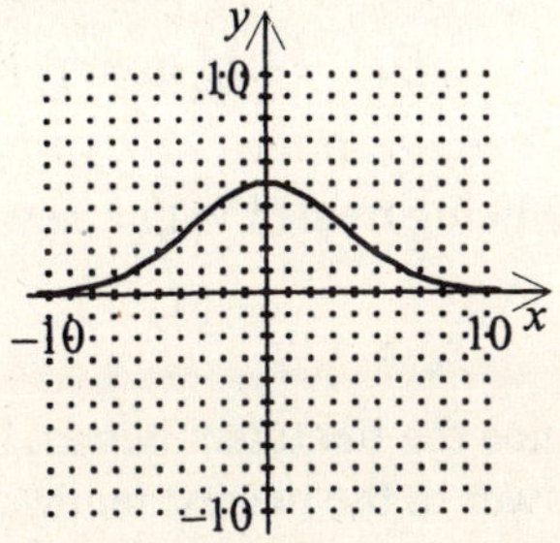

(A) Exponential decay model (B) Exponential growth model

(C) Logarithmic model (D) Gaussian model

66. Which equation represents the logarithmic model of a function?

(A) $y = \dfrac{4}{1 + 3e^{-0.98x}}$ (B) $y = 3 + 4\ln x$ (C) $y = 3e^{-(x-5)^2/4}$ (D) $y = 3e^{-0.98x}$

67. Sketch the graph of $y = e^{-0.75x}$.

68. Identify the model shown in the graph.

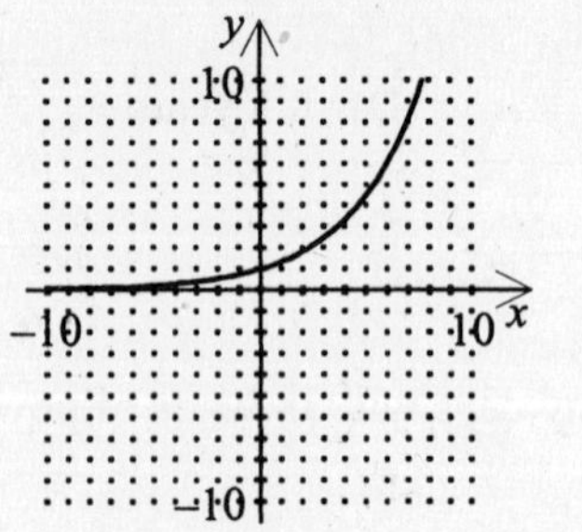

Objective 2: Use exponential growth and decay functions to model and solve real-life problems

69. The number of bison per acre of range in the wild is
$N = 2.4 \times 10^{5-0.003w}$
where N is the number of bison per acre and w is the average weight of the bison in pounds. Find the average weight of a bison in a herd that has an average of nine animals per acre of range.

(A) 1623 lb (B) 1328 lb (C) 1475 lb (D) 1425 lb

70. The number of bacteria N in a culture is modeled by
$N = 150e^{kt}$
where t is the time in hours. If $N = 270$ when $t = 6$, what is the time required for the original population to quadruple in size?

(A) 24.00 hours (B) 22.97 hours (C) 16.33 hours (D) 14.15 hours

71. The population of a bacteria culture with an initial population of 6500 being treated with a new antibiotic can be modeled by
$N = 6500e^{-0.2t}$
where N is the number of bacteria present and t is the time in hours since the treatment began. In how many hours will the culture have a count of 2925? Round the answer to the nearest tenth.

72. A new car with a purchase price of $24,000 has a value of $12,000 three years later.
(a) Write the straight-line model $V = mt + b$.
(b) Write the exponential model $V = ae^{kt}$.
(c) Find the book values of the car after four years using each model.
(d) Interpret the meaning of the slope in the straight-line model.

Objective 3: Use Gaussian functions to model and solve real-life problems

73. The life expectancy of a fluorescent tube follows a normal distribution

$y = 0.001e^{-(x-4000)^2/320{,}000}$

where x is the life expectancy in hours. Use a graphing utility to sketch the function and find the average life of a fluorescent tube.

(A) 3000 hr (B) 4000 hr (C) 4500 hr (D) 3200 hr

74. The ages of the employees of a local company roughly follow a normal distribution

$y = 0.133e^{-(x-36)^2/18}$

where x is the age of the employee. Use a graphing utility to sketch the function and find the average age of an employee.

(A) 38 (B) 18 (C) 34 (D) 36

75. During boot camp, a drill sergeant measured the weight of the men in his unit. He found the weights roughly followed a normal distribution

$y = 0.0249e^{-(x-157)^2/512}$

where x is the weight of a recruit in pounds. Use a graphing utility to sketch the function and find the average weight of the men.

76. The heights of 1000 girls at Ocean View High School were recorded and found to follow the normal distribution

$y = 0.1995e^{-(x-65)^2/8}$

where x is the height of a female student in inches. Use a graphing utility to sketch the function and find the average height of the girls at Ocean View High School.

Objective 4: Use logistic growth functions to model and solve real-life problems

77. A virus is accidentally brought to a remote village with a population of 5000 that has never been exposed to the disease. The spread of the virus is modeled by

$$y = \frac{5000}{1+4999e^{-0.6t}}$$

where t is the time in days since the virus was introduced. How many villagers will be infected after 10 days?

(A) 336 (B) 373 (C) 429 (D) 354

78. A rumor begins at a closed-campus high school that the principal has donated one million dollars for the graduation party. There are 2700 students in the high school and the spread of the rumor is modeled by

$$y = \frac{2700}{1 + 2699e^{-0.3t}}$$

where y is the number of students who have heard the rumor and t is the time in minutes. How long before 80% of the students have heard the rumor?

(A) 35.6 min (B) 29.4 min (C) 31.0 min (D) 27.9 min

79. Bacteria growing in a petri dish is modeled by

$$y = \frac{577{,}000}{1 + 65.2e^{-0.46t}}$$

where t is the time in hours. Find the number of bacteria after 4 hours.

80. A new product was heavily advertised for several months, but then advertising was sharply reduced. The peak sales were 100,000 units but are now decreasing and can be modeled by

$$S = \frac{100{,}000}{1 + 0.6e^{0.05t}}$$

where t is the time in months. In how many months will the sales be 50% of the peak sales? Round the answer to the nearest tenth.

Objective 5: Use logarithmic functions to model and solve real-life problems

81. Chemists measure the acidity of an aqueous solution using pH. The pH is

$$\text{pH} = -\log_{10}\left[\text{H}^+\right]$$

where pH is a measure of the hydrogen ion concentration $\left[\text{H}^+\right]$ measured in moles of hydrogen per liter of a solution. Find the pH of a solution if $\left[\text{H}^+\right] = 8.5 \times 10^{-7}$.

(A) 6.1 (B) 0.9 (C) 7.9 (D) 13.1

82. The level of sound β in decibels with an intensity of I is

$$\beta = 10 \log_{10} \frac{I}{I_0}$$

where I_0 is 10^{-12} watt/m^2, corresponding roughly to the faintest sound that can be heard by the human ear. Find the level of sound β of a whistling teakettle with $I = 10^{-4}$ watt/m^2.

(A) 80 decibels (B) 160 decibels (C) 75 decibels (D) 90 decibels

83. Due to the installation of a highway sound barrier, the noise level near a busy highway was reduced from 110 decibels to 101 decibels. Find the percent decrease, to the nearest tenth of one percent, in the intensity level of the noise as a result of the sound barrier.

84. Chemists measure the acidity of an aqueous solution using pH. The pH is

$$pH = -\log_{10}\left[H^{+}\right]$$

where pH is a measure of the hydrogen ion concentration $\left[H^{+}\right]$ measured in moles of hydrogen per liter of a solution. Compute $\left[H^{+}\right]$ for a solution in which $pH = 8.40$.

Answer Key for Chapter 5 Exponential and Logarithmic Functions

Section 5.1: Exponential Functions and Their Graphs

Objective 1: Recognize and evaluate exponential functions with base a

[1] (D)

[2] (A)

[3] 12.126

[4] 0.894

Objective 2: Graph exponential functions

[5] (B)

[6] (A)

[7] 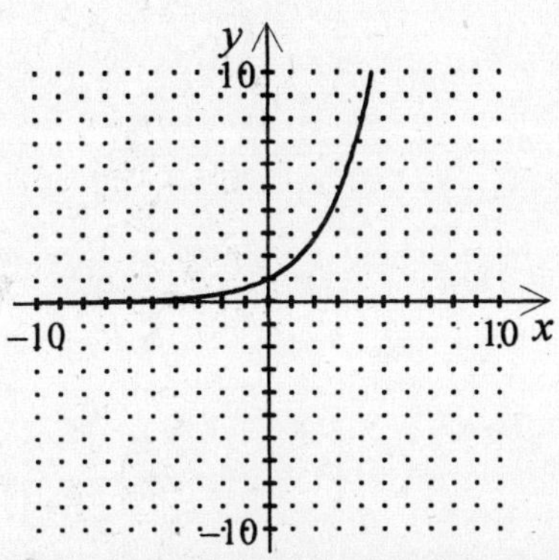

[8] 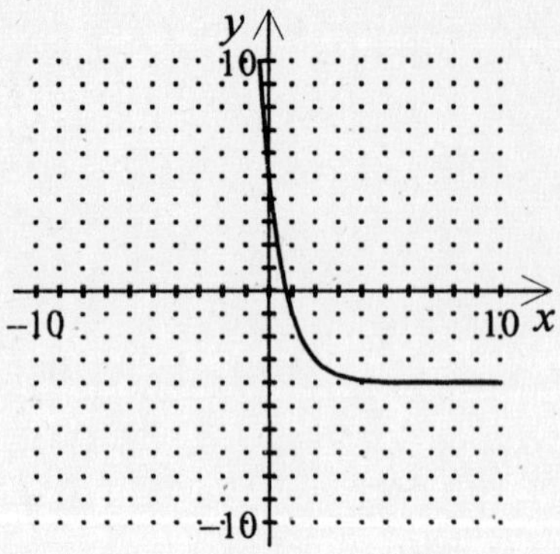

Objective 3: Recognize and evaluate exponential functions with base e

[9] (A)

[10] (A)

[11] 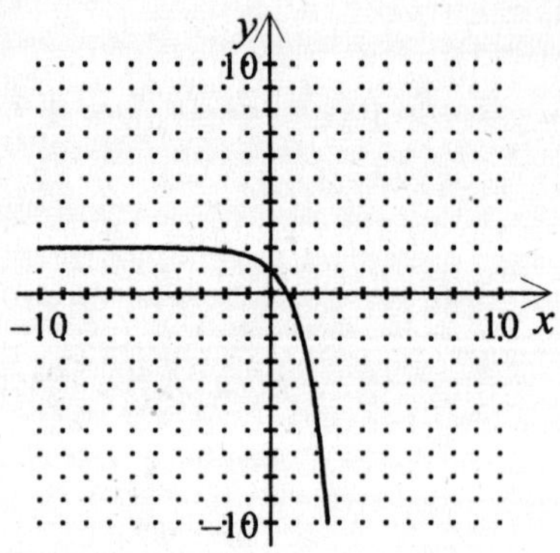

[12] 1.237

Objective 4: Use exponential functions to model and solve real-life applications

[13] (D)

[14] (C)

[15] 27.8 g

[16] 3.5 lb/in^2

Section 5.2: Logarithmic Functions and Their Graphs

Objective 1: Recognize and evaluate logarithmic functions with base a

[17] (B)

[18] (D)

[19] $\log_2 32 = 5$

[20] –5

Objective 2: Graph logarithmic functions

[21] (C)

[22] (A)

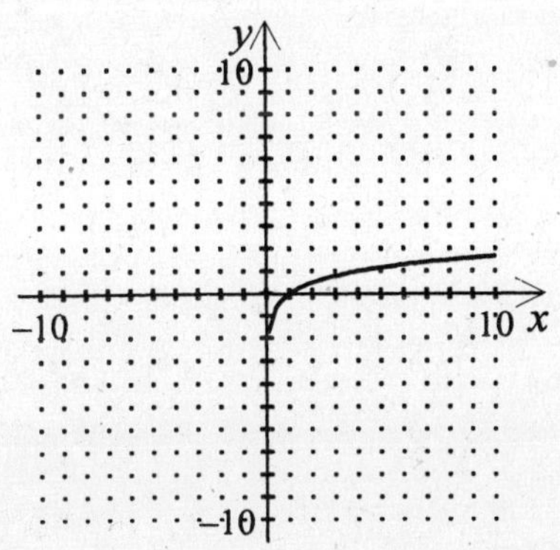

[23]

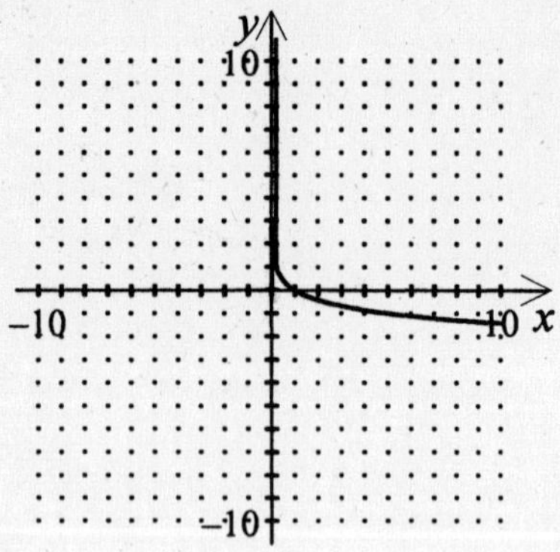

[24] ____________________________________

Objective 3: Recognize and evaluate natural logarithmic functions

[25] (D)

[26] (B)

[27] 1.907

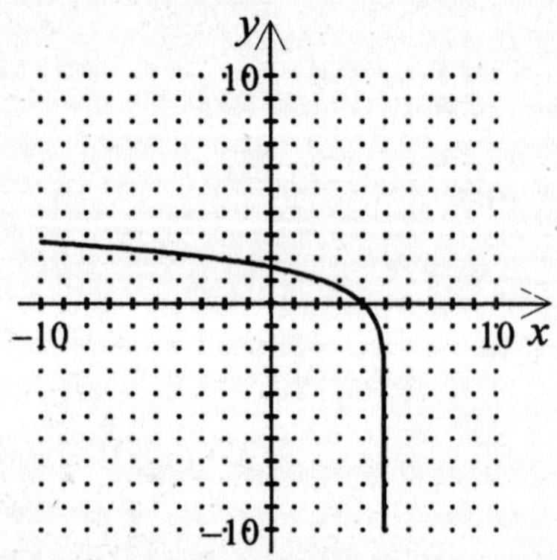

[28] ____________________________________

Objective 4: Use logarithmic functions to model and solve real-life applications

[29] (C)

[30] (A)

[31] 6.198 decibels

[32] Magnitude of 7.0

Section 5.3: Properties of Logarithms

Objective 1: Use the change-of-base formula to rewrite and evaluate logarithmic expressions

[33] (A)

[34] (A)

[35] 1.796

$$f(x)=\frac{\log_{10} x}{\log_{10} \frac{5}{3}}$$

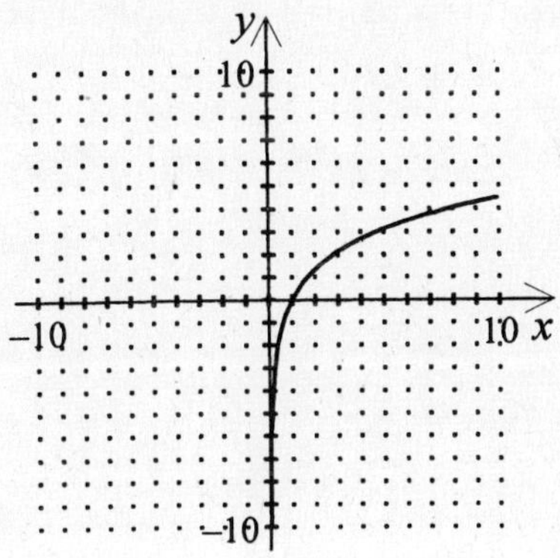

[36]

Objective 2: Use properties of logarithms to evaluate or rewrite logarithmic expressions

[37] (B)

[38] (B)

[39] False. Answers may vary. Sample answer: $\log_{10}(6\cdot 7)=\log_{10} 6+\log_{10} 7$.

[40] 3

Objective 3: Use properties of logarithms to expand or condense logarithmic expressions

[41] (D)

[42] (C)

[43] $\log_6 \dfrac{(x+6)^6(x+4)^2}{\sqrt{x}}$

[44] $\log_a 4 + \log_a x + 4\log_a y - 3\log_a z$

Objective 4: Use logarithmic functions to model and solve real-life applications

[45] (D)

[46] (C)

[47] (a) $10(\log_{10} P_o - \log_{10} P_i) = \log_{10}\left(\dfrac{P_o}{P_i}\right)^{10}$
(b) 15.584 dB

[48] (a) $E = \log\left(\dfrac{C_2}{C_1}\right)^{1.4}$
(b) -0.843 kcal/mol

Section 5.4: Exponential and Logarithmic Equations

Objective 1: Solve simple exponential and logarithmic equations

[49] (D)

[50] (C)

[51] $\ln\dfrac{1}{4} \approx -1.386$

[52] 1000

Objective 2: Solve more complicated exponential equations

[53] (C)

[54] (B)

[55] 1.262, 1.771

[56] 283.321

Objective 3: Solve more complicated logarithmic equations

[57] (D)

[58] (C)

[59] $x = 4$

[60] 4.525

Objective 4: Use exponential and logarithmic equations to model and solve real-life applications

[61] (B)

[62] (A)

[63] 82 days

[64] 4

Section 5.5: Exponential and Logarithmic Models

Objective 1: Recognize the five most common types of models involving exponential and logarithmic functions

[65] (D)

[66] (B)

[67] 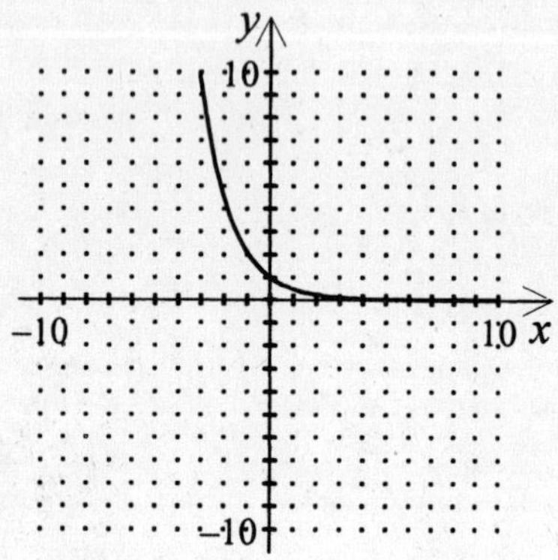

[68] Exponential growth model

Objective 2: Use exponential growth and decay functions to model and solve real-life problems

[69] (C)

[70] (D)

[71] 4.0 hr

[72]
(a) $V = -4000t + 24{,}000$
(b) $V = 24{,}000e^{-0.231t}$
(c) Straight-line, \$8000; Exponential, \$9524
(d) The slope represents the yearly decrease in value.

Objective 3: Use Gaussian functions to model and solve real-life problems

[73] (B)

[74] (D)

[75] 157 lb

[76] 65 in.

Objective 4: Use logistic growth functions to model and solve real-life problems

[77] (B)

[78] (C)

[79] 50,815

[80] 10.2 months

Objective 5: Use logarithmic functions to model and solve real-life problems

[81] (A)

[82] (A)

[83] 87.4%

[84] 3.98×10^{-9} mol/L

Chapter 6 Topics in Analytic Geometry

Section 6.1: Lines

Objective 1: Find the inclination of a line

1. Find, in radians and degrees, the inclination θ of the line passing through the points.

   $(-4, -2), (3, 1)$

   (A) 2.74 radians, 156.8° (B) 0.29 radian, 20°

   (C) 2.85 radians, 160° (D) 0.4 radian, 23.2°

2. Find, in radians and degrees, the inclination θ of the line.

   $-5x - 13y - 66 = 0$

   (A) 0.26 radian, 24.2° (B) 2.77 radians, 159°

   (C) 2.88 radians, 155.8° (D) 0.37 radian, 21°

3. Find, in radians and degrees, the inclination θ of the line with a slope of m.

   $m = \frac{5}{7}$

4. Find, in radians and degrees, the inclination θ of the line passing through the points.

   $(-5, -6), (-7, 8)$

Objective 2: Find the angle between two lines

Find, in radians and degrees, the angle θ between the lines.

5. $12x - 10y + 9 = 0$

$8x + 2y - 1 = 0$

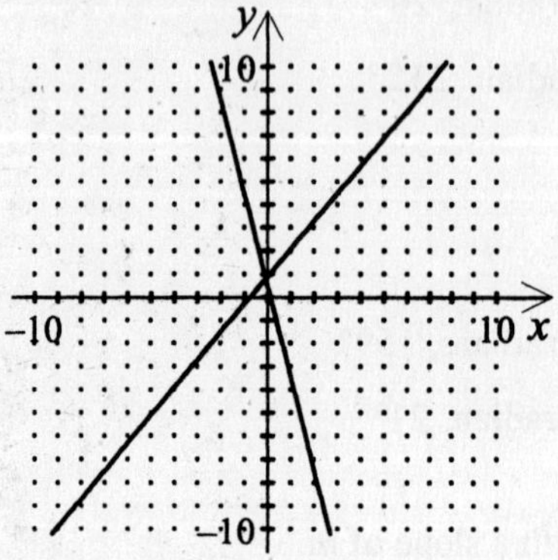

(A) 0.94 radian, 53.8°

(B) 2.09 radians, 129.4°

(C) 2.2 radians, 126.2°

(D) 1.05 radians, 50.6°

6. $0.9x + 0.6y - 0.2 = 0$

$1.1x + 0.3y + 1.0 = 0$

(A) 0.32 radian, 18.4°

(B) 0.43 radian, 15.2°

(C) 2.71 radians, 164.8°

(D) 2.82 radians, 161.6°

7. $-x + 0.3y + 0.6 = 0$

$-0.9x - 0.8y - 1.2 = 0$

Find, in radians and degrees, the angle θ between the lines.

8. $9x - 3y - 6 = 0$

$-x + 4y + 5 = 0$

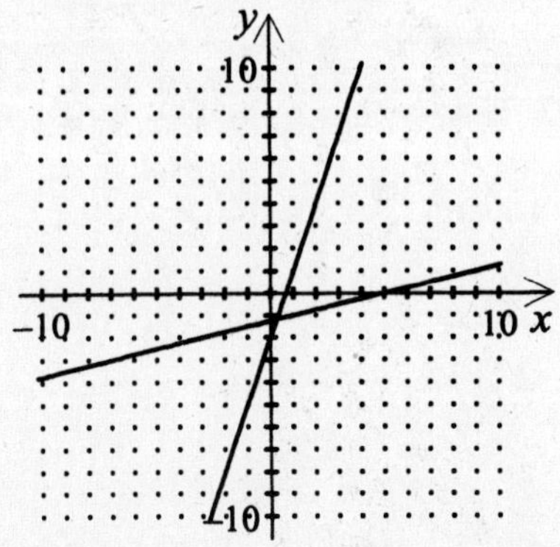

Objective 3: Find the distance between a point and a line

9. Find the distance between the parallel lines.

$5x + 7y = 1$

$5x + 7y = -14$

(A) 0.46 (B) 1.74 (C) 8.6 (D) 8.66

10. Find the distance between the point and the line.

$(0,\ 0)\quad 7x + 4y = -2$

(A) 13 (B) 8.31 (C) 0.35 (D) 0.25

11. Find the distance from $(-6, -7)$ to the line containing $(3, -6)$ and $(6, -3)$, to the nearest hundredth.

12. Find (a) the altitude from vertex B of the triangle to side AC, and (b) the area of the triangle.

$A = (-11,\ -17),\ B = (-14,\ 8),\ C = (-13,\ -15)$

Section 6.2: Introduction to Conics: Parabolas

Objective 1: Recognize a conic as the intersection of a plane and a double-napped cone

13. Identify the diagram that represents a circle.

(A)

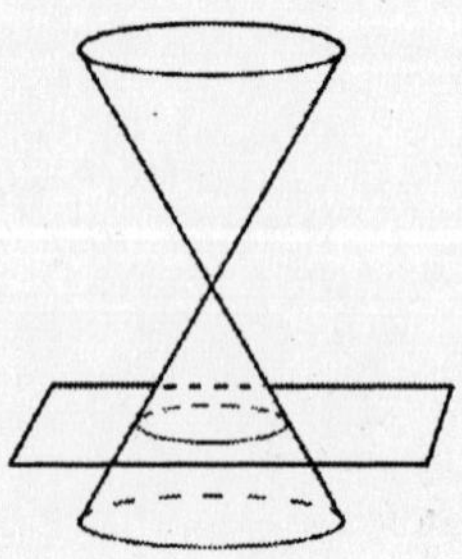

(B)

(C)

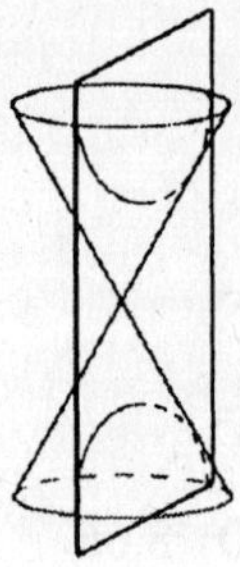

(D)

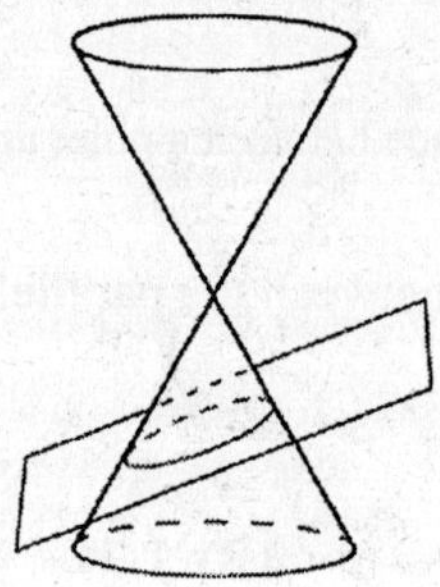

14. Identify the conic.

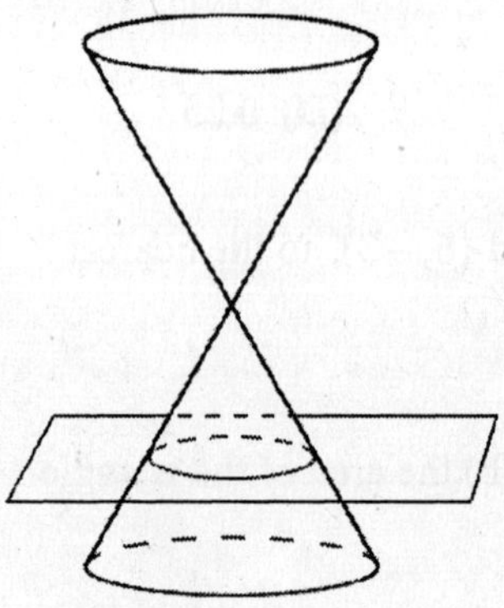

(A) Ellipse (B) Hyperbola (C) Parabola (D) Circle

15. Draw a plane on the following diagram so that the intersection of the plane with the double-napped cone is a hyperbola.

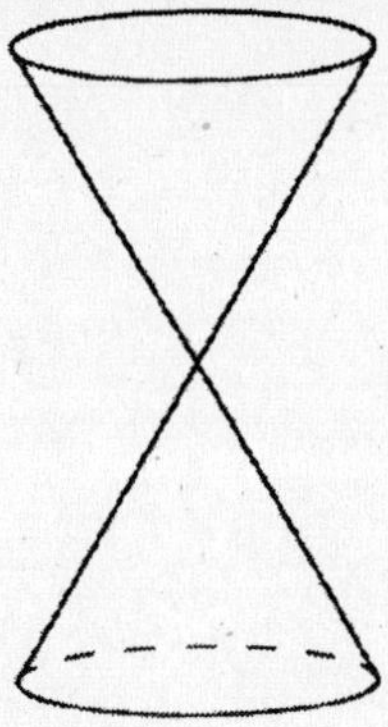

16. Draw a plane on the picture of the double-napped cone so that the intersection of the plane and the cone matches the conic section depicted on the left.

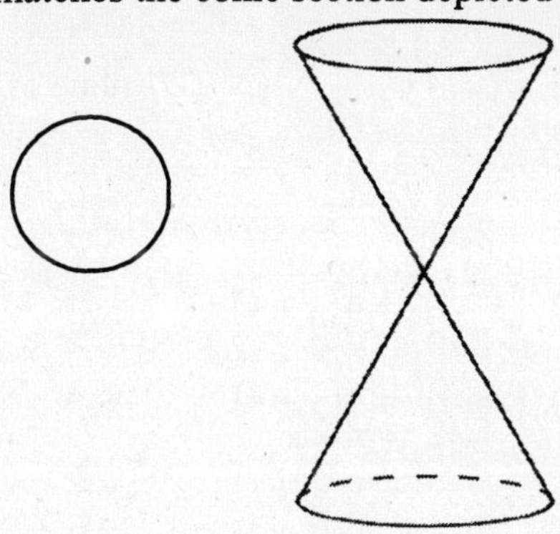

Objective 2: Write the standard form of the equation of a parabola

Find the standard form of the equation of the parabola.

17.

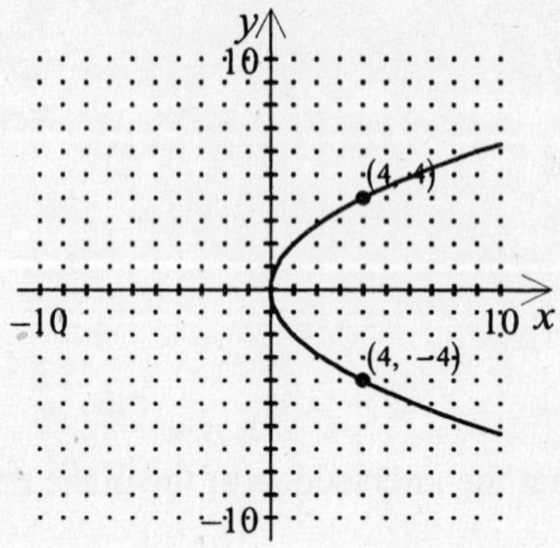

(A) $x^2 = 4y$ (B) $y^2 = -4x$ (C) $y^2 = 4x$ (D) none of these

18. Focus: $(5, 1)$; Directrix: $x = 3$

(A) $(x-1)^2 = 4(y-4)$ (B) $(y+1)^2 = 4(x+4)$

(C) $(y-1)^2 = 4(x-4)$ (D) $(x+1)^2 = 4(y+4)$

19.

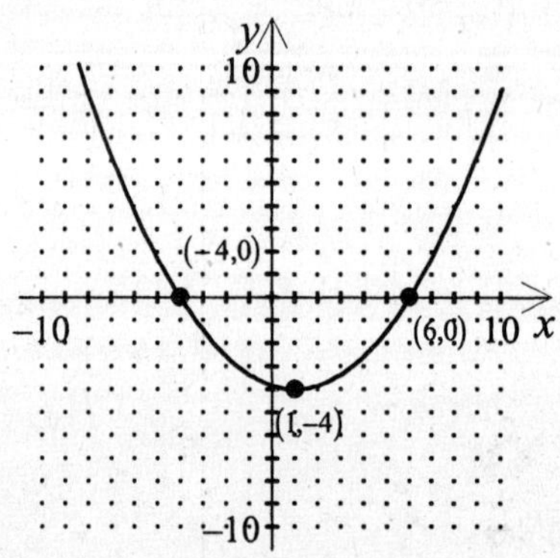

20. Vertex: $(0, 0)$; Focus: at $(0, -6)$

Objective 3: Use the reflective property of parabolas to solve real-life problems

21. In a factory, a parabolic mirror to be used in a searchlight was placed on the floor. It measured 50 centimeters tall and 70 centimeters wide. Find an equation of the parabola with its vertex at the origin.

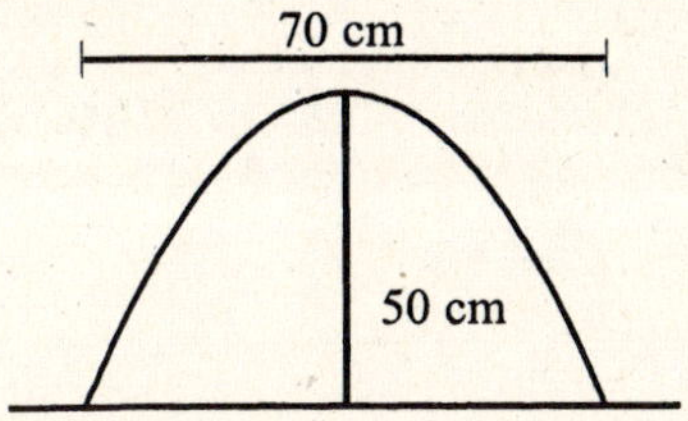

(A) $y=-\frac{2}{49}x^2$ (B) $y=-\frac{1}{98}x^2$ (C) $y=\frac{2}{49}x^2$ (D) $y=\frac{1}{98}x^2$

22. A parachute maker has been experimenting with different parabolic shapes. A particular parachute that is 23.5 feet wide is 9.8 feet higher in the center than it is on the edge (see figure).

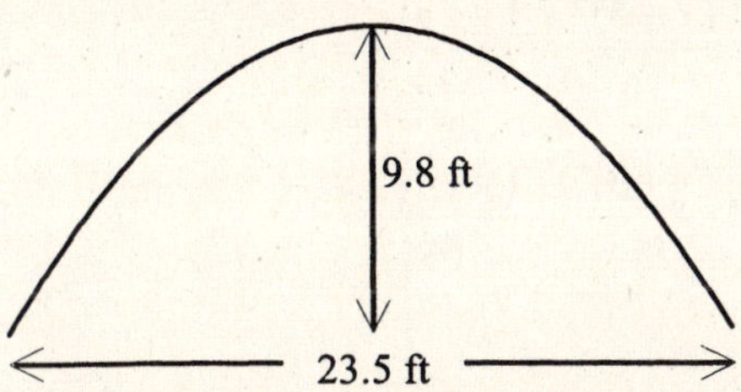

(a) Find an equation of the parabola that models the curve of the parachute. (Assume that the origin is at the center of the parachute.)
(b) How far from the center of the parachute is the parachute surface 3 feet lower than in the middle? (Round the answer to the nearest hundredth.)

(A) (a) $y=\frac{-196}{11{,}045}x^2$
(b) 6.50 ft

(B) (a) $y=\frac{-2350}{2401}x^2$
(b) 7.15 ft

(C) (a) $y=\frac{-784}{11{,}045}x^2$
(b) 6.50 ft

(D) (a) $y=\frac{-3136}{11{,}045}x^2$
(b) 7.15 ft

23. Each cable of a suspension bridge is suspended in the shape of a parabola between two towers that are 149 meters apart and whose tops are 29 meters above the roadway. The cables are 2 meters above the roadway midway between the towers.
(a) Find an equation for the parabolic shape of the cable. (Assume the origin is on the road, midway between the towers.)
(b) Find the height of the cable 6 meters from the center of the bridge. (Round the answer to the nearest thousandth.)

24. A parabolic television dish antenna is 15 feet across at its opening and 9 feet deep. If the receiver is placed at the focus, how far is it from the vertex?

Section 6.3 Ellipses

Objective 1: Write the standard form of the equation of an ellipse

Find the standard form of the equation of the specified ellipse.

25. Vertices: $(0,\ \pm 3)$; Foci: $(\pm 4,\ 0)$

(A) $\frac{x^2}{25}+\frac{y^2}{9}=4$ (B) $\frac{x^2}{9}+\frac{y^2}{25}=4$ (C) $\frac{x^2}{25}+\frac{y^2}{9}=1$ (D) $\frac{x^2}{9}+\frac{y^2}{25}=1$

26. Foci: $(2,\ -3)$, $(-4,\ -3)$; Major axis of length 8

(A) $\frac{(x+1)^2}{7}+\frac{(y+3)^2}{16}=1$ (B) $\frac{(x+3)^2}{16}+\frac{(y+1)^2}{7}=1$

(C) $\frac{(x+1)^2}{16}+\frac{(y+3)^2}{7}=1$ (D) $\frac{(x+3)^2}{7}+\frac{(y+1)^2}{16}=1$

27. Center: $(5,\ 2)$; Vertex: $(14,\ 2)$; Minor axis of length 8

28. Center: $(0,\ 0)$; Foci: $\left(\pm 4\sqrt{2},\ 0\right)$, Vertices: $(\pm 9,\ 0)$

Objective 2: Use properties of ellipses to model and solve real-life problems

29. Lena rides her bike through a tunnel shaped like the top half of an ellipse. The tunnel is 9.8 meters wide and 2.7 meters high. On her bike, the top of Lena's helmet is 1.8 meters above the ground. If she were to ride through the tunnel 3.3 meters from the center, would her helmet miss the ceiling? If so, by how much?

(A) Yes, she would clear the ceiling by 0.4 m.

(B) Yes, she would clear the ceiling by 0.2 m.

(C) Yes, she would clear the ceiling by 0.3 m. (D) No, she would hit the ceiling.

30. The opening under a bridge is in the shape of a semielliptical arch and has a span of 102 feet. The height of the arch, at a distance of 24 feet from the center, is 29 feet. Find the height of the arch at its center.

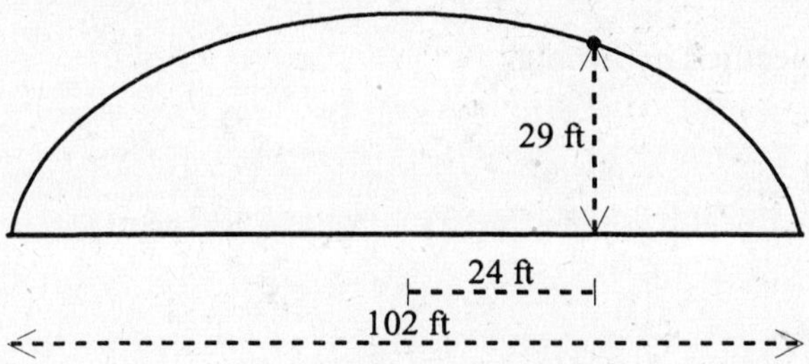

(A) $32\frac{2}{3}$ ft (B) $33\frac{13}{15}$ ft (C) $33\frac{1}{2}$ ft (D) $32\frac{13}{15}$ ft

31. A satellite orbits around the moon in an elliptical path. Assuming the moon to be the center of a rectangular coordinate system, find the equation of the elliptical path of a satellite whose x-intercepts are $\pm 111{,}000$ km and whose y-intercepts are $\pm 59{,}000$ km.

32. A skating park has a track shaped like an ellipse. If the length of the track is 80 m and the width of the track is 42 m, find the equation of the ellipse.

Objective 3: Find the eccentricity of an ellipse

Find the eccentricity of the ellipse.

33. $16x^2 + 25y^2 - 160x - 200y + 400 = 0$ (A) $\frac{\sqrt{2}}{3}$ (B) $\frac{5\sqrt{41}}{41}$ (C) $\frac{\sqrt{41}}{5}$ (D) $\frac{3}{5}$

34. $\frac{x^2}{49} + \frac{y^2}{25} = 1$ (A) $\frac{2\sqrt{6}}{7}$ (B) $\frac{5\sqrt{3}}{7}$ (C) $\frac{3\sqrt{3}}{4}$ (D) $\frac{2}{5}$

35. $25x^2 - 250x + 36y^2 - 360y + 625 = 0$

36. $\frac{(x-2)^2}{64} + \frac{(y-2)^2}{9} = 1$

Section 6.4: Hyperbolas

Objective 1: Write the standard form of the equation of a hyperbola

Find the standard form of the equation of the specified hyperbola.

37.

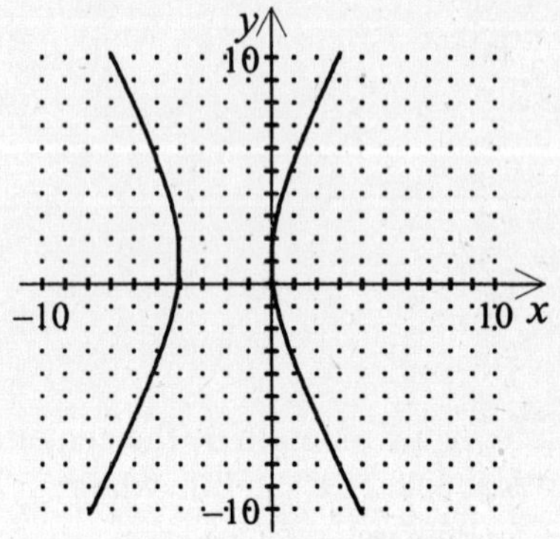

(A) $\dfrac{(x-2)^2}{16}-\dfrac{(y+1)^2}{4}=1$ (B) $\dfrac{(x+1)^2}{16}-\dfrac{(y-2)^2}{4}=1$

(C) $\dfrac{(x-1)^2}{4}-\dfrac{(y+2)^2}{16}=1$ (D) $\dfrac{(x+2)^2}{4}-\dfrac{(y-1)^2}{16}=1$

38. Center: $(-5, 4)$
Focus: $(25, 4)$
Vertex: $(-23, 4)$

(A) $\dfrac{(x-5)^2}{324}-\dfrac{(y+4)^2}{576}=1$ (B) $\dfrac{(x+5)^2}{324}-\dfrac{(y-4)^2}{576}=1$

(C) $\dfrac{(x+5)^2}{324}+\dfrac{(y-4)^2}{576}=1$ (D) $\dfrac{(y-4)^2}{324}-\dfrac{(x+5)^2}{576}=1$

39. $9x^2-108x-36y^2-72y-36=0$

40. Vertices: $(\pm 2, 0)$

Asymptotes: $y=\pm\dfrac{3}{2}x$

Objective 2: Find the asymptotes of a hyperbola

41. Find the equations of the asymptotes of the hyperbola.
$-4x^2 + 25y^2 + 8x + 100y - 4 = 0$

(A) Asymptotes: $y = \pm \frac{5}{2}x$ (B) Asymptotes: $y = \pm \frac{2}{5}x$

(C) Asymptotes: $y = \frac{2}{5}x - \frac{12}{5},\ y = -\frac{2}{5}x - \frac{8}{5}$ (D) Asymptotes: $y = \frac{5}{2}x - \frac{9}{2},\ y = -\frac{5}{2}x + \frac{1}{2}$

42. Find the vertices, foci, and the equations of the asymptotes of the hyperbola.
$\frac{x^2}{25} - \frac{y^2}{36} = 1$

(A) Vertices: $(0,\ \pm 5)$; Foci: $(0,\ \pm\sqrt{61})$; Asymptotes: $y = \pm \frac{6}{5}x$

(B) Vertices: $(\pm 5,\ 0)$; Foci: $(\pm\sqrt{61},\ 0)$; Asymptotes: $y = \pm \frac{6}{5}x$

(C) Vertices: $\left(0,\ \pm\frac{25}{36}\right)$; Foci: $(\pm\sqrt{61},\ 0)$; Asymptotes: $x = \pm \frac{5}{6}y$

(D) Vertices: $\left(\pm\frac{25}{36},\ 0\right)$; Foci: $(0,\ \pm\sqrt{61})$; Asymptotes: $x = \pm \frac{5}{6}y$

Find the equations of the asymptotes of the hyperbola.

43. $(x+2)^2 - \frac{9(y-4)^2}{25} = 1$

44. $25y^2 - 16x^2 - 400 = 0$

Objective 3: Use properties of hyperbolas to solve real-life problems

45. You and a friend live 2 miles apart and are talking on the phone. You hear a crack of thunder and 3 seconds later your friend hears the crack. Find an equation that gives the possible places where the lightning could have occurred. (Use miles as the unit of distance. The speed of sound is about 1100 ft/sec.)

(A) $\dfrac{x^2}{\frac{25}{256}} - \dfrac{y^2}{\frac{231}{256}} = 1$ (B) $\dfrac{x^2}{\frac{231}{256}} - \dfrac{y^2}{\frac{25}{256}} = 1$ (C) $\dfrac{x^2}{\frac{25}{256}} + \dfrac{y^2}{\frac{231}{256}} = 1$ (D) $\dfrac{x^2}{\frac{231}{256}} + \dfrac{y^2}{\frac{25}{256}} = 1$

46. The focus of a hyperbolic mirror has coordinates $(30, 0)$. Find the vertex of the mirror if $(30, -24)$ is one of the points on the mirror and the other focus has coordinates $(-30, 0)$.

(A) $(0, 18)$ (B) $(18, 0)$ (C) $(6, 0)$ (D) $(54, 0)$

47. An explosion is monitored by three scientific stations. Station A is located at $(4000, 0)$, Station B at $(-4000, 0)$, and Station C at $(-4000, -6600)$. Station A records the explosion 3 seconds after Station B. Station C records the explosion 6 seconds after Station B. If the coordinates are measured in feet and sound travels at 1100 feet per second, determine the coordinates of the explosion.

48. Two LORAN stations are positioned 204 miles apart along a straight shore. The stations transmit synchronized pulses which travel at the speed of light (186,000 miles per second) in order to find the coordinates of ships in the area. The station closest to a ship being tracked by the LORAN stations is called the master station.
(a) A ship records a time difference of 0.000516 second between the LORAN signals. Determine where the ship would reach shore if it were to follow the indicated hyperbolic path.
(b) If the ship were to enter a harbor located between the two stations 12 miles from the master station, what time difference should it be looking for?
(c) If the ship is 80 miles offshore when the desired time difference is obtained, what is the exact location of the ship?

Objective 4: Classify a conic from its general equation

49. Identify the equation of a parabola.

(A) $4x^2 - 8x + 4y^2 - 6y - 6 = 0$ (B) $2y^2 - 8y - 6x^2 + 4x - 6 = 0$

(C) $-6x^2 - 8x - 12y^2 - 6y = -8$ (D) $-6y^2 - 8x - 6y = 2$

50. Identify the equations that represent parabolas.

(i) $2y^2 + 2x + 2y + 13 = 0$

(ii) $28x^2 + 4x + 8y + 15 = 0$

(iii) $3y^2 + 3x + 11y + 2 = 0$

(A) i and iii only (B) ii only (C) i, ii, and iii (D) i and ii only

Classify the graph of the equation as a circle, a parabola, an ellipse, or a hyperbola.

51. $3x^2 - 10x - 2y - 3 = 0$

52. $4x^2 - 7y^2 + 10x - 3 = 0$

Section 6.5: Rotation of Conics

Objective 1: Rotate the coordinate axes to eliminate the xy-term in the equation of a conic

53. The $x'y'$ -coordinate system has been rotated θ degrees from the *xy*-coordinate system. The coordinates of a point on the *xy*-coordinate system are given. Find the coordinates of the point on the rotated coordinate system.

$\theta = 35°$, $(4, 6)$

(A) $x' = -0.16$, $y' = 7.21$ (B) $x' = 5.57$, $y' = -1.47$

(C) $x' = 2.62$, $y' = 6.72$ (D) $x' = 6.72$, $y' = -2.62$

54. Rotate the axes to eliminate the *xy*-term.

$4x^2 + 14xy + 4y^2 + 28x - 28y + 37 = 0$

(A) $11x^2 - 3y^2 - 28\sqrt{2}y + 37 = 0$ (B) $11x^2 - 3y^2 - 28\sqrt{2}x + 37 = 0$

(C) $-3y^2 - 28\sqrt{2}x - 28\sqrt{2}y + 37 = 0$ (D) $11x^2 - 28\sqrt{2}x - 28\sqrt{2}y + 37 = 0$

55. Rotate the axes to eliminate the *xy*-term. Sketch the graph of the resulting equation, showing both sets of axes.

$xy - 2 = 0$

56. Graph the conic. Determine the angle θ through which the axes are rotated.

$6x^2 - 60xy - 19y^2 = 1014$

Objective 2: Use the discriminant to classify a conic

Find the discriminant and classify the graph.

57. $8x^2 + 13xy - 5y^2 + 3x - 3y = 0$

(A) 329; ellipse (B) 329; hyperbola (C) 9; parabola (D) 9; ellipse

58. $x^2 - 2xy + y^2 - 5 = 0$

(A) 8; ellipse (B) 8; hyperbola (C) 0; ellipse (D) 0; parabola

Use the discriminant to classify the graph, use the Quadratic Formula to solve for y, and then use a graphing utility to graph the equation.

59. $5x^2 - 9xy + 5y^2 + 5x + 3y = 0$

60. $5x^2 + 2xy + 5y^2 - 4x + 5y - 61 = 0$

Section 6.6: Parametric Equations

Objective 1: Evaluate a set of parametric equations for a given value of the parameter

Evaluate the set of parametric equations for the given value of the parameter.

61. $x = 3t$, $y = t^2 + 2t$, $t = 2$

(A) $x = 6, y = 40$ (B) $x = 6, y = 8$ (C) $x = 2, y = 8$ (D) $x = 2, y = 40$

62. $x = 3 - 3t$, $y = 4t^5$, $t = 4$

(A) $x = 0, y = 80$ (B) $x = 0, y = 4096$ (C) $x = -9, y = 4096$ (D) $x = -9, y = 16$

63. $x = -1 - 2\cos\theta$, $y = -3 + 2\sin\theta$, $\theta = \dfrac{3\pi}{4}$

64. $x = 2t^3$, $y = -3\ln t$, $t = 3$

Objective 2: Sketch the curve that is represented by a set of parametric equations

65. Sketch the curve represented by the parametric equations (indicate the direction of the curve).

$x = \tan t$

$y = \sec t$

(A)

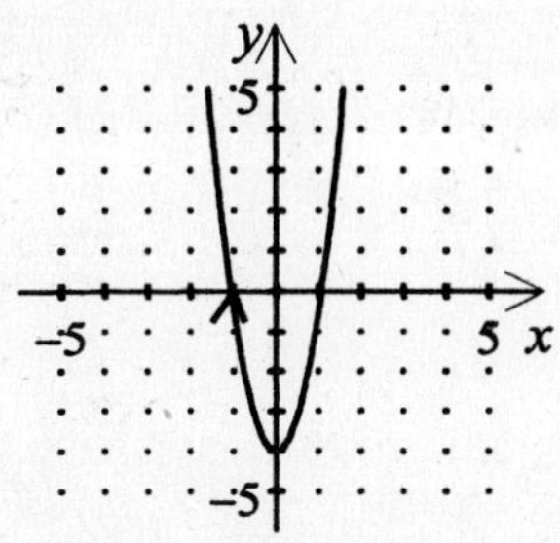

(B)

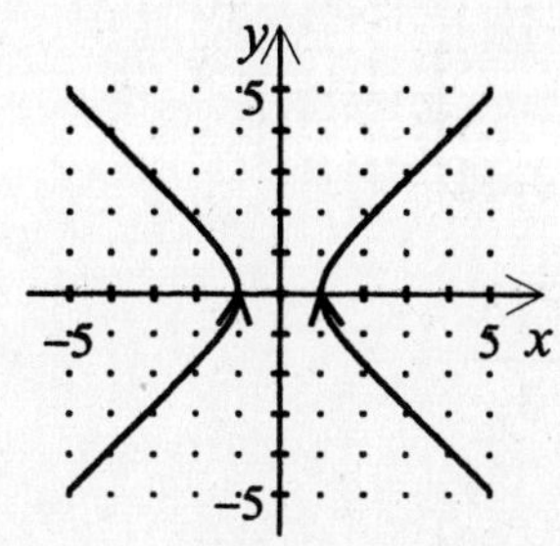

(C)

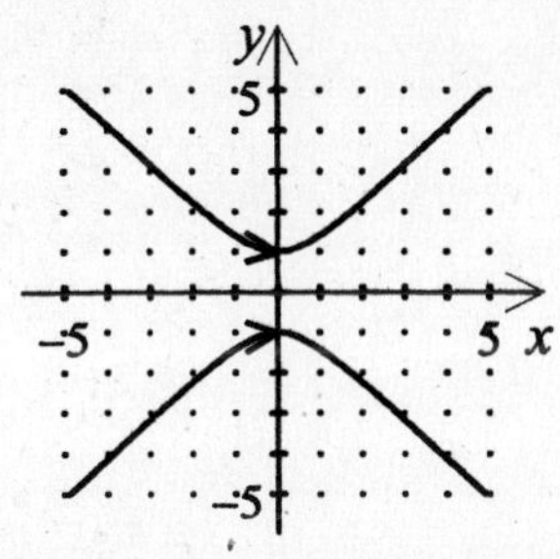

(D)

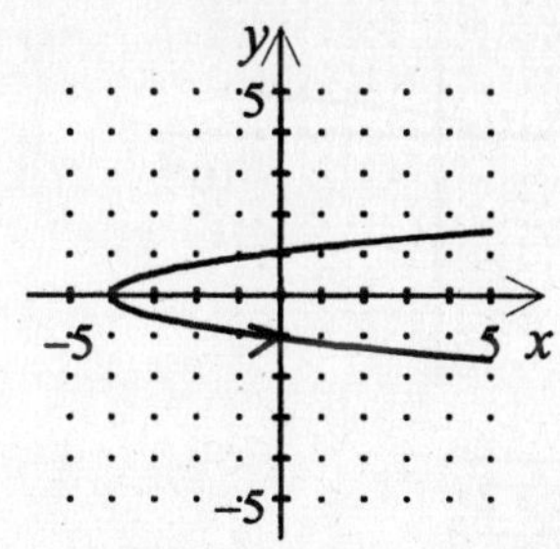

Sketch the curve represented by the parametric equations.

66. $x = 3t - 2$

$y = \sqrt{t}$

(A) $y = 2x^2 + 3$

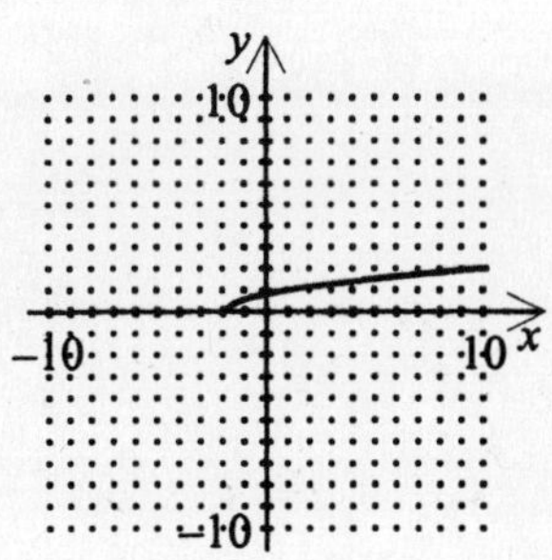

(B) $y = 2x^2 + 3$

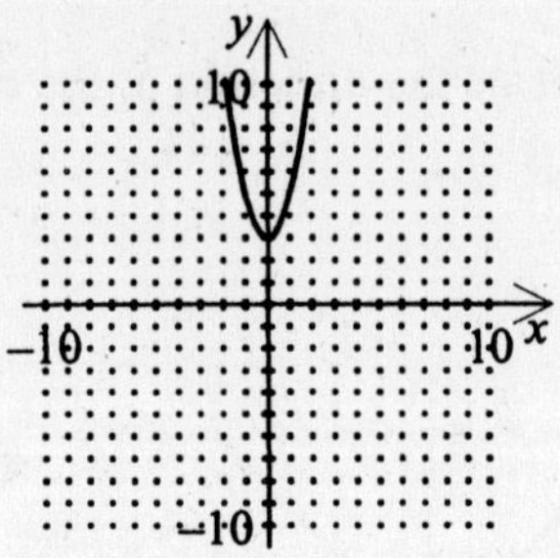

(C) $y = \sqrt{\dfrac{x+2}{3}}$

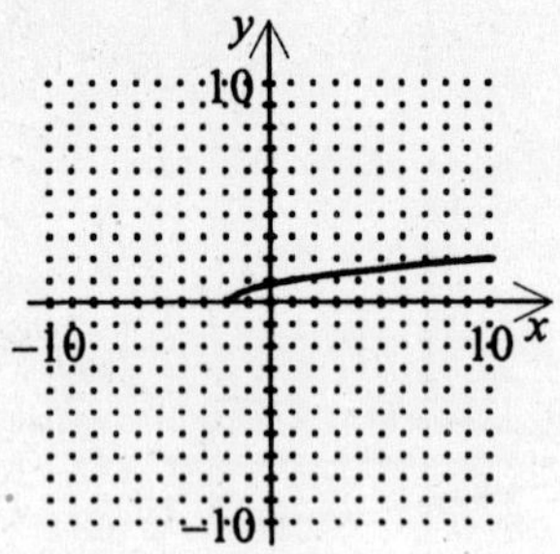

(D) $y = \sqrt{\dfrac{x+2}{3}}$

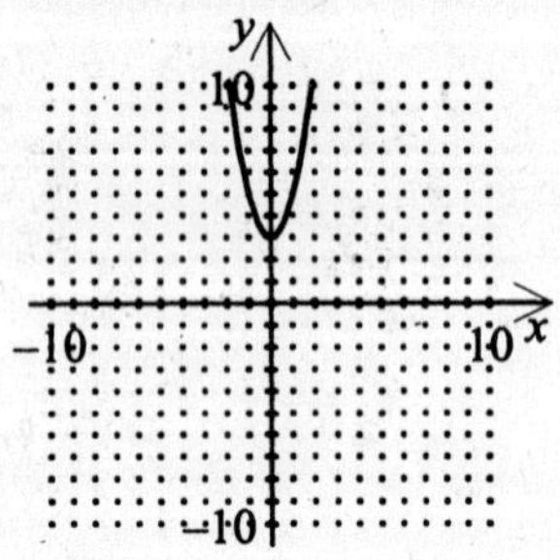

(66.)

Sketch the curve represented by the parametric equations.

67. $x = 2\sin 2t$

$y = 5\cos 2t$

68. Sketch the curve represented by the parametric equations (indicate the direction of the curve).

$x = \tan t$

$y = -\sec t$

Objective 3: Rewrite a set of parametric equations as a single rectangular equation

Eliminate the parameter and obtain the standard form of the rectangular equation.

69. $x = -8t + 5$

$y = -2t + 2$

(A) $y = \frac{1}{4}x + \frac{3}{4}$ (B) $y = -\frac{1}{4}x - \frac{3}{4}$ (C) $y = \frac{1}{2}x - \frac{3}{2}$ (D) $y = 8x - \frac{4}{3}$

70. $x = t$

$y = t^2 + 3t - 2$

(A) $y = \sqrt{x^2 + 3x - 2}$ (B) $x^2 + 2x - 2 = 0$ (C) $t = x$ (D) $y = x^2 + 3x - 2$

71. $x = 5\sec t$

$y = 4\tan t$

72. $x = 4t^2 + 5$

$y = 7t - 3$

Objective 4: Find a set of parametric equations for a graph

Identify the set of parametric equations for the given rectangular equation.

73. $x^2 + y^2 = 121$

(A) $x = 11\cos t$, $y = 11\sin t$ (B) $x = 11\cot t$, $y = 11\tan t$ (C) $x = 11t$, $y = 11t$ (D) $x = \sqrt{t}$, $y = 11$

74. $y = x^2 - 3x$

(A) $x = x^3,\ y = x^2 - 3x$ (B) $x = t^2 - 3t,\ y = t$

(C) $x = 0,\ y = x^2 - 3x$ (D) $x = t,\ y = t^2 - 3t$

75. Find a set of parametric equations for the given rectangular equation using $t = \dfrac{x}{3}$.

$$y = -\frac{1}{9}x^2 - x + 10$$

76. Find a set of parametric equations for the given rectangular equation using $t = x$.

$$y = x^3 + x + 2$$

Section 6.7: Polar Coordinates

Objective 1: Plot points on the polar coordinate system

77. Identify the graph of the given point.

$$\left(6, \frac{17\pi}{12}\right)$$

(A)

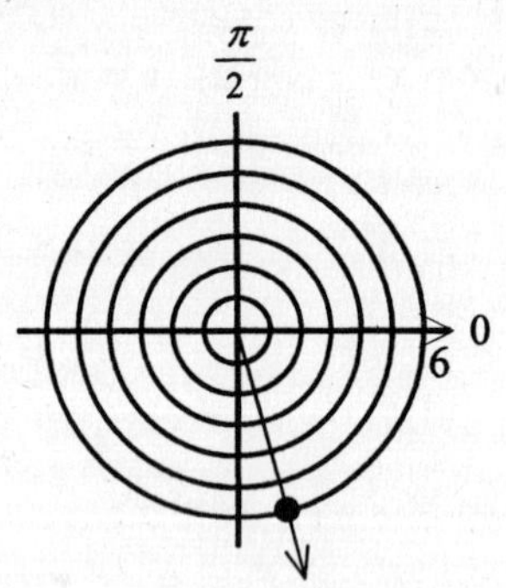

(B)

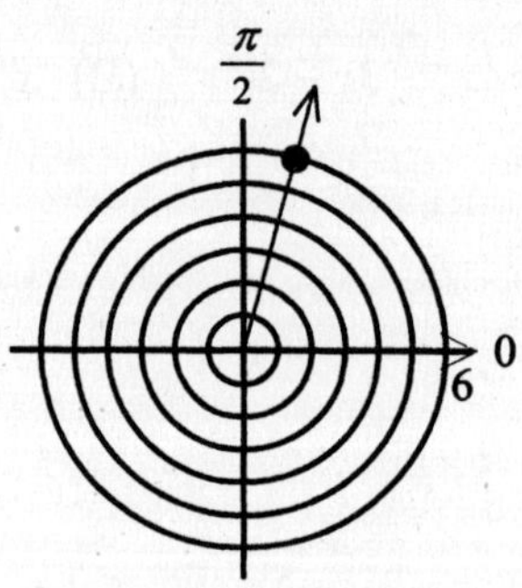

(C)

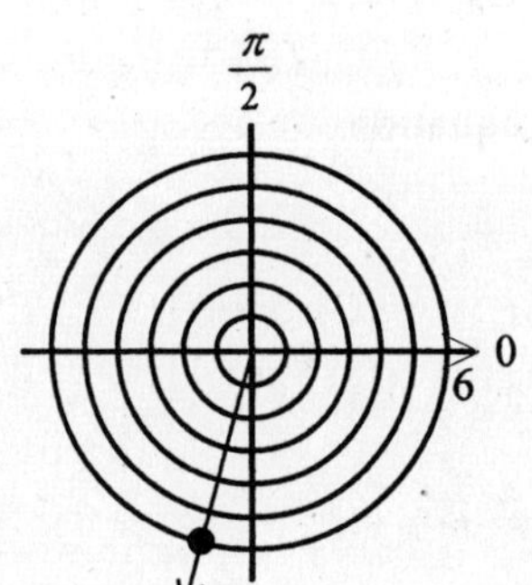

(D)

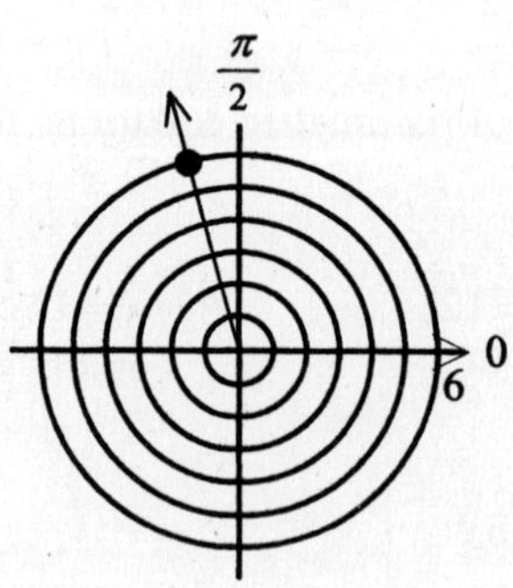

78. Identify the point graphed.

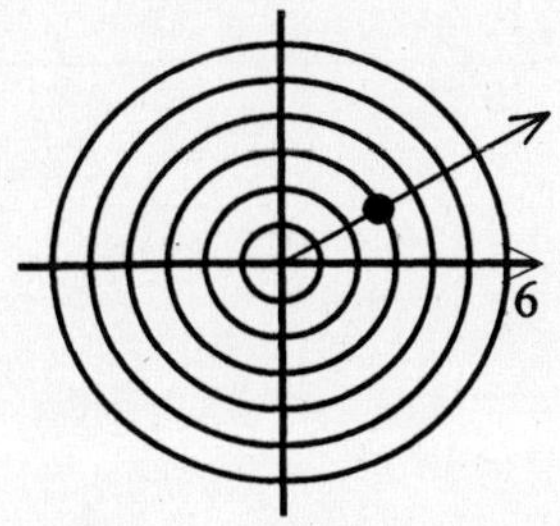

(A) $\left(3, \frac{\pi}{3}\right)$ (B) $\left(3, \frac{\pi}{6}\right)$ (C) $\left(-3, \frac{\pi}{6}\right)$ (D) $\left(-3, \frac{\pi}{3}\right)$

Plot the point given in polar coordinates and find two additional polar representations.

79. $\left(4, -\frac{\pi}{6}\right)$

80. $(4, -1.28)$

Objective 2: Convert points from rectangular to polar form and vice versa

81. A point in rectangular coordinates is given. Convert the point to polar coordinates.
$(3, -3)$

(A) $\left(3\sqrt{2}, \frac{9\pi}{4}\right)$ (B) $\left(-3\sqrt{2}, \frac{7\pi}{4}\right)$ (C) $\left(3\sqrt{2}, \frac{7\pi}{4}\right)$ (D) $\left(3\sqrt{2}, \frac{5\pi}{4}\right)$

82. A point in polar coordinates is given. Convert the point to rectangular coordinates.
$(5, 2.13)$

(A) $(-2.653, 4.238)$ (B) $(4.997, 0.186)$

(C) $(-0.531, 0.848)$ (D) $(-0.186, -4.997)$

83. A point in rectangular coordinates is given. Convert the point to polar coordinates.
$\left(-2, 2\sqrt{3}\right)$

84. A point in polar coordinates is given. Convert the point to rectangular coordinates.
$(9, \pi)$

Objective 3: Convert equations from rectangular to polar form and vice versa

85. Convert the polar equation to rectangular form.
$r = -12 \cos \theta$

(A) $x^2 + (y-6)^2 = 36$ (B) $(x+6)^2 + y^2 = 36$

(C) $x^2 + (y+6)^2 = 36$ (D) $(x-6)^2 + y^2 = 36$

Convert the rectangular equation to polar form.

86. $x^2 + (y+8)^2 = 64$

(A) $r = \pm 8$ (B) $r = 16 \sin \theta$ (C) $r = -16 \cos \theta$ (D) $r = -16 \sin \theta$

87. $x^2 + (y-2)^2 = 4$

88. Convert the polar equation to rectangular form.
$$r = \frac{3}{7 - 2\cos\theta}$$

Section 6.8: Graphs of Polar Equations

Objective 1: Graph polar equations by point plotting

Identify the graph of the polar equation.

89. $r = 5 \sin 3\theta, \ 0 \le \theta \le 2\pi$

(A)

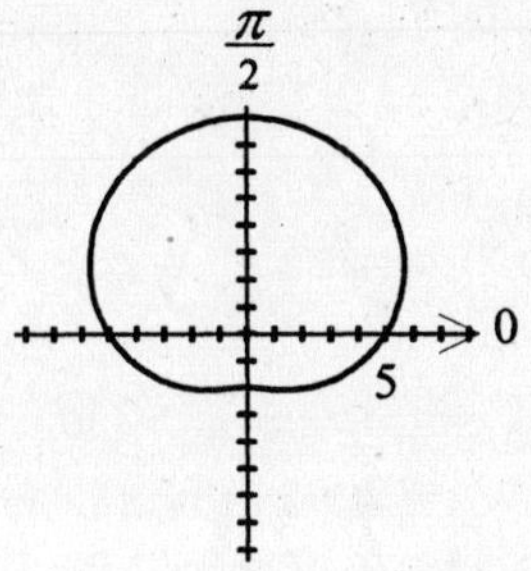

(B)

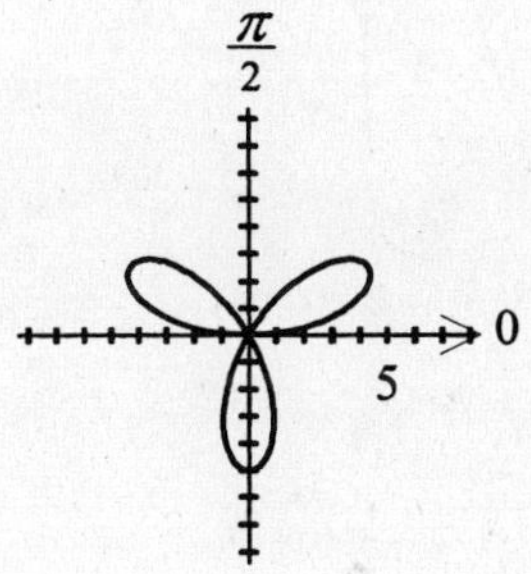

(C)

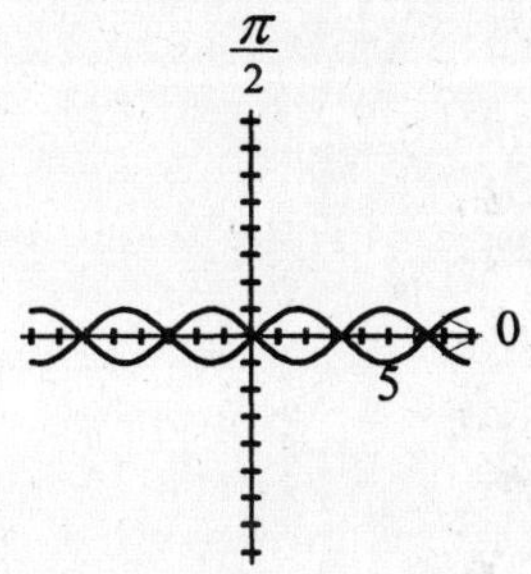

(D)

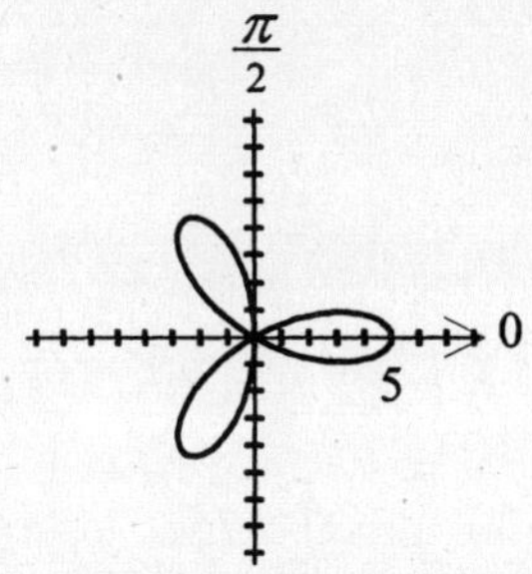

(89.)

Identify the graph of the polar equation.

90. $2 = r\cos(\theta - 20°)$

(A)

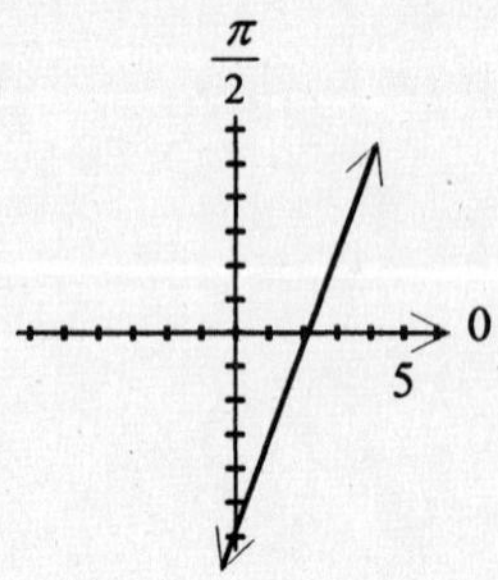

(B)

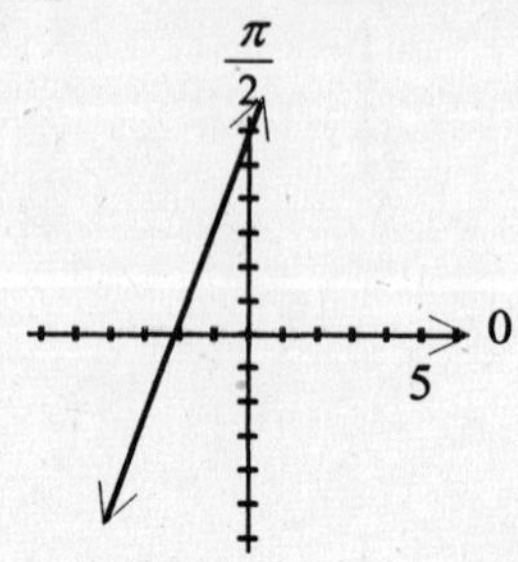

(C)

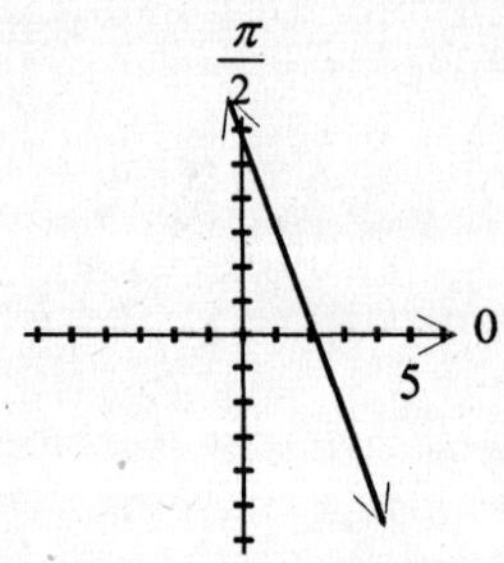

(D)

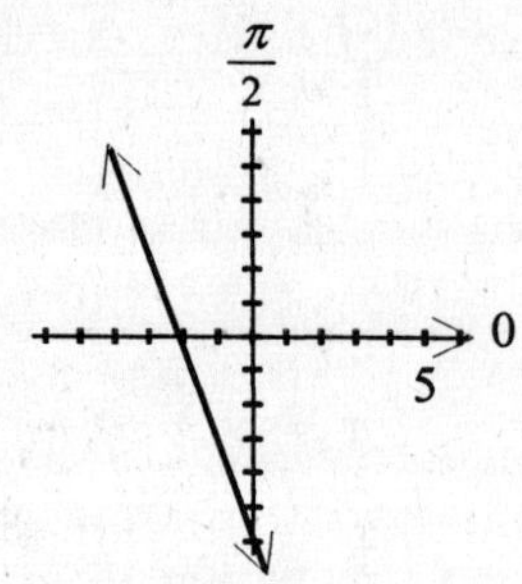

Sketch the graph of the polar equation.

91. $r = \dfrac{16}{5 - 3\cos\theta}$

92. $\theta = \dfrac{5\pi}{3}$

Objective 2: Use symmetry to sketch graphs of polar equations

93. Test for symmetry with respect to $\theta = \frac{\pi}{2}$, the polar axis, and the pole.

$r = \frac{1}{1 - 2\cos\theta}$

(A) The line does not have symmetry.

(B) The graph is symmetric with respect to $\theta = \frac{\pi}{2}$.

(C) The graph is symmetric with respect to the pole.

(D) The graph is symmetric with respect to the polar axis.

94. Test for symmetry with respect to $\theta = \frac{\pi}{2}$, the polar axis, and the pole.

$r = \sin 3\theta$

(A) The graph is symmetric with respect to the polar axis.

(B) The graph is symmetric with respect to the line $\theta = \frac{\pi}{2}$ and the pole.

(C) The graph is symmetric with respect to the polar axis and the pole.

(D) The graph is symmetric with respect to the line $\theta = \frac{\pi}{2}$.

95. Test for symmetry with respect to $\theta = \frac{\pi}{2}$, the polar axis, and the pole.

$r = \frac{1}{3 - \sin\theta}$

96. Test for symmetry with respect to $\theta = \frac{\pi}{2}$, the polar axis, and the pole.

$r = -2\cos 4\theta$

Objective 3: Use zeros and maximum r-values to sketch graphs of polar equations

97. Find the maximum value of $|r|$ and any zeros of r.

$r = 10(1 + 2\sin\theta)$

(A) 10; $\theta = \frac{\pi}{6}, \frac{11\pi}{6}$

(B) 10; $\theta = \frac{\pi}{6}, \frac{5\pi}{6}$

(C) 30; $\theta = \frac{5\pi}{6}, \frac{7\pi}{6}$

(D) 30; $\theta = \frac{7\pi}{6}, \frac{11\pi}{6}$

98. Sketch the graph of the polar equation $r = \dfrac{9}{5 + 4\cos\theta}$.

(A)

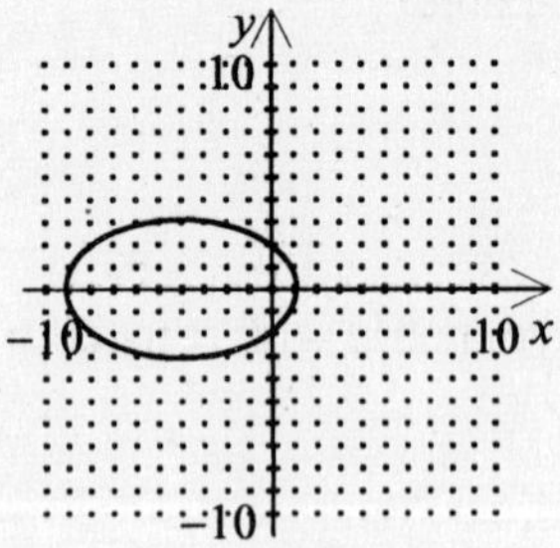

(B)

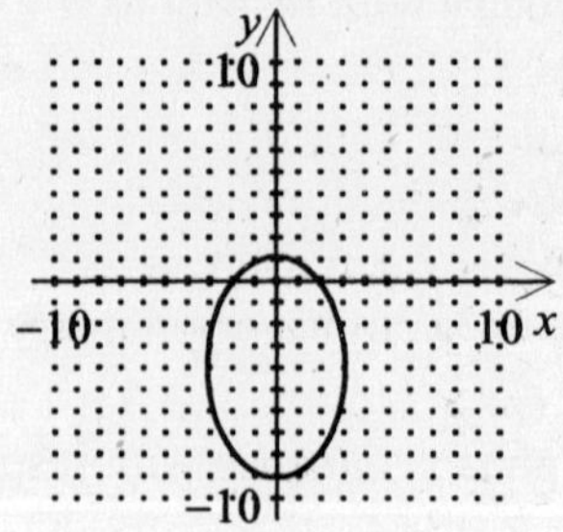

(C)

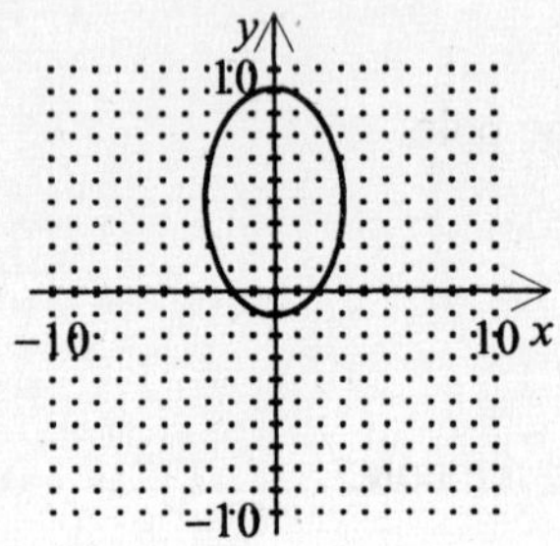

(D)

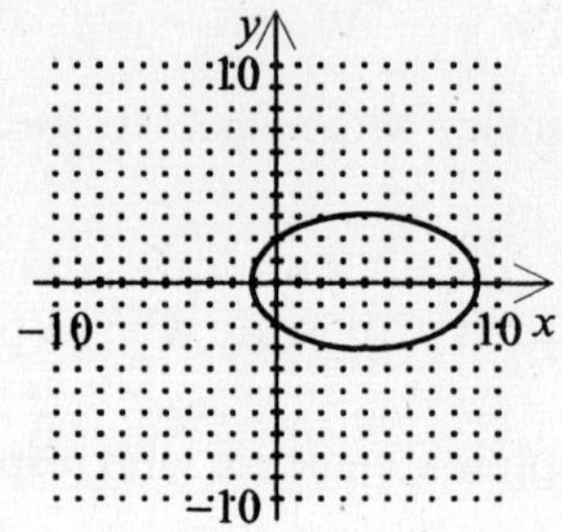

99. Find the maximum value of $|r|$ and any zeros of r.

$r = -8 \sin 4\theta$

100. Find the maximum value of $|r|$ and any zeros of r.

$r = 6\left(\sqrt{3} + 2\cos\theta\right)$

Objective 4: Recognize special polar graphs

101. Identify the type of polar graph.

$r^2 = 4\cos 2\theta$

(A)

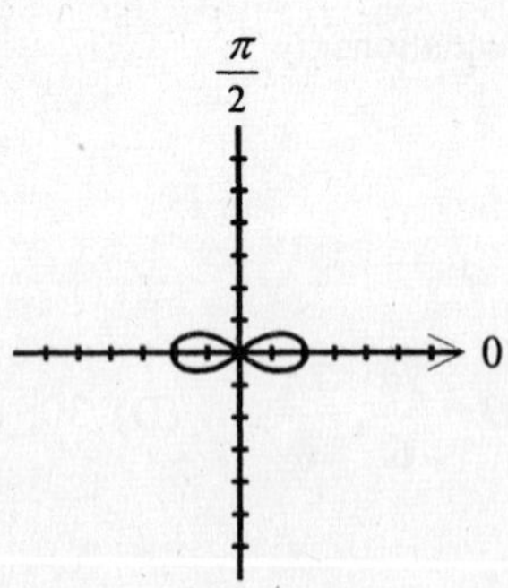

(B)

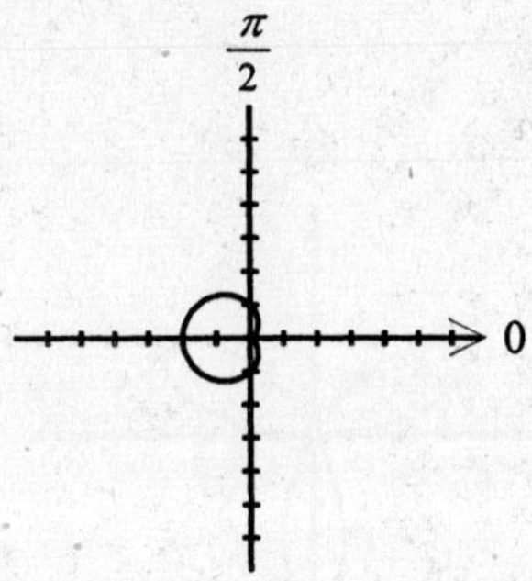

(C)

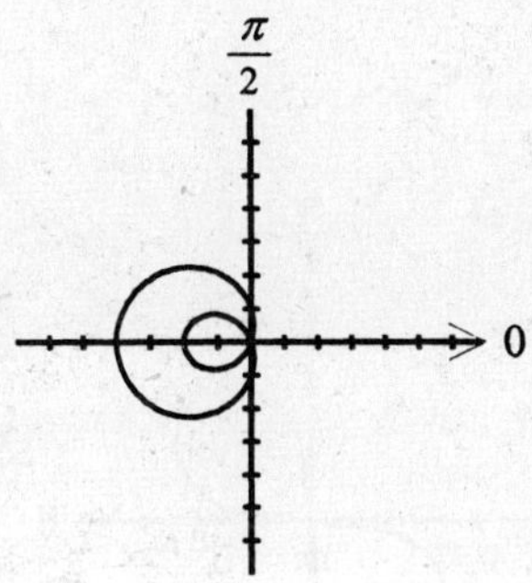

(D)

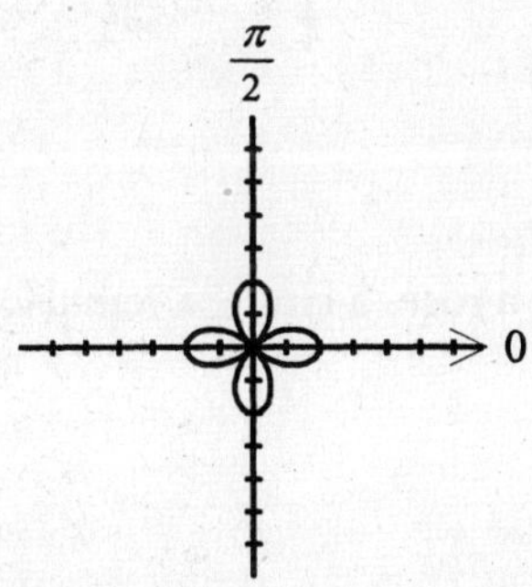

(101.)

102. Identify the graph of a lemniscate.

(A)

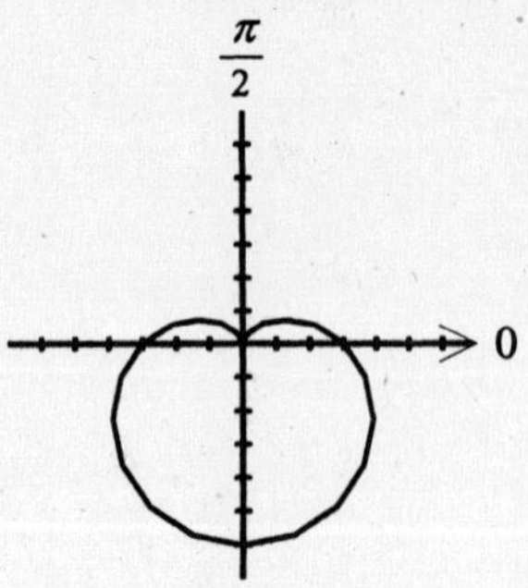

(B)

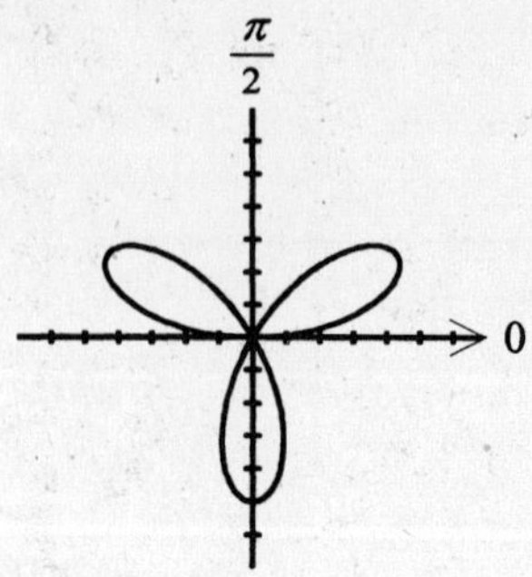

(C)

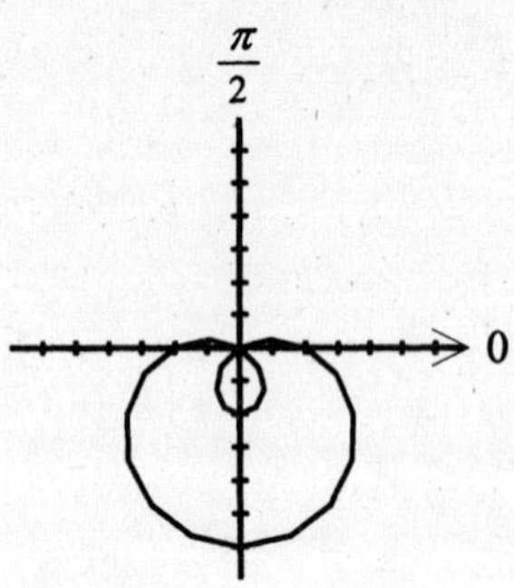

(D)

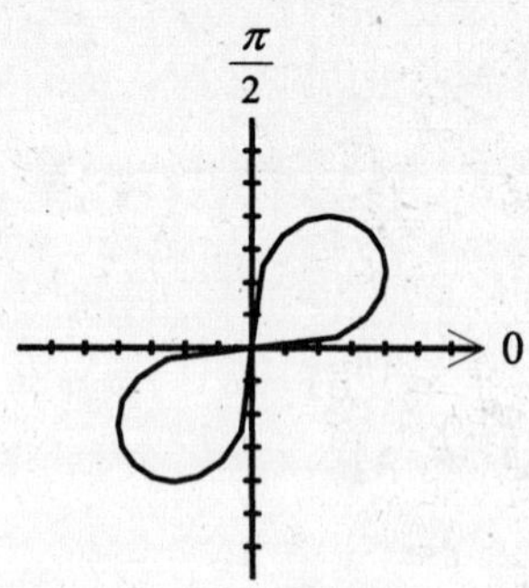

103. Identify the graph for the equation $r = 2\sin\theta$, as a cardioid, a rose, a circle, a lemniscate, or a limaçon.

104. Identify the type of polar graph.
$r = 6\cos 4\theta$

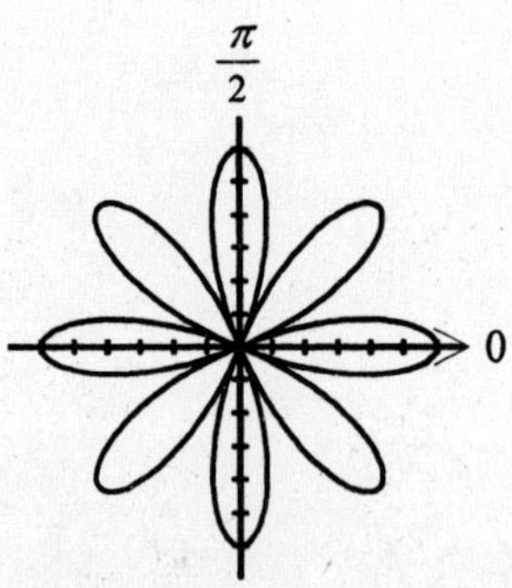

Section 6.9: Polar Equations of Conics

Objective 1: Define conics in terms of eccentricity

105. Graph.

$$r = \frac{2}{3 + 4\sin\theta}$$

(A)

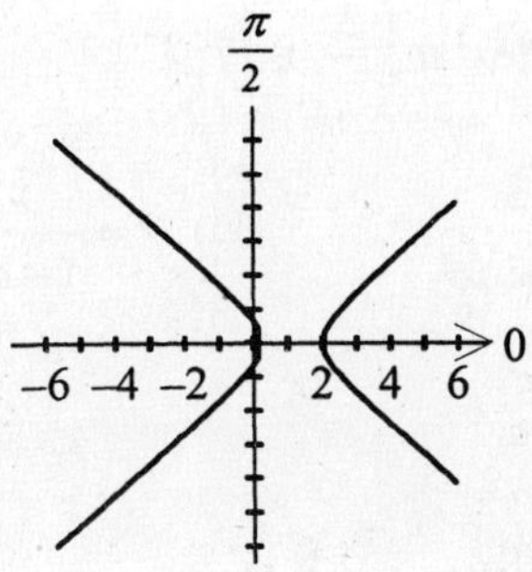

(B)

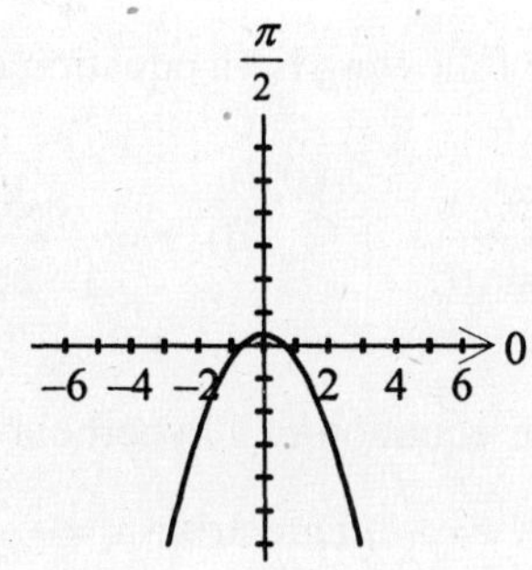

(C)

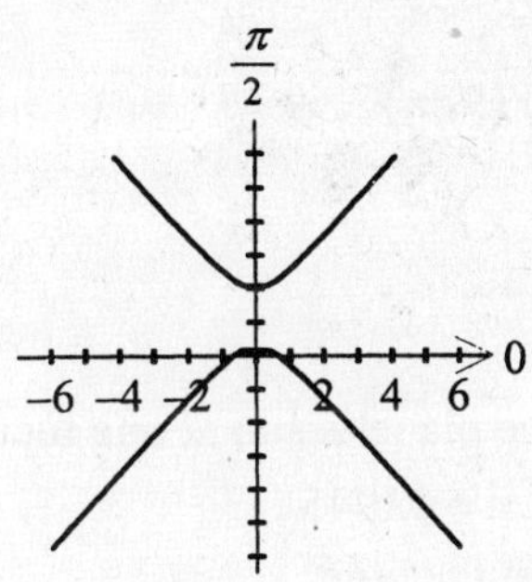

(D)

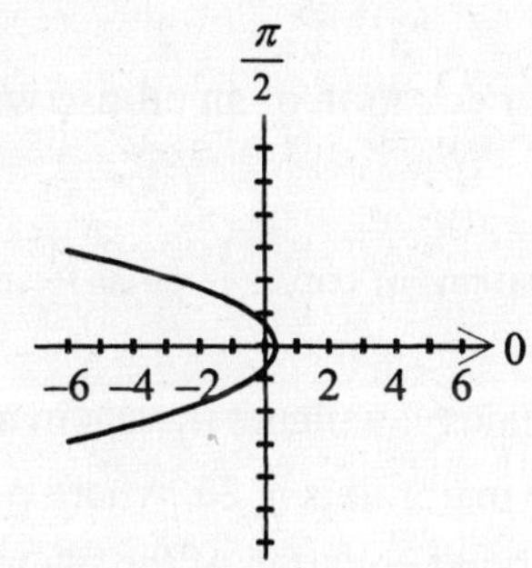

106. Identify the equation of an ellipse.

(A) $r = \frac{5}{2 - 3\cos\theta}$ (B) $r = \frac{5}{2 - 3\sin\theta}$ (C) $r = \frac{5}{3 - 2\cos\theta}$ (D) $r = \frac{5}{2 - 2\cos\theta}$

107. Use a graphing utility to graph the polar equation. Identify the graph.

$$r = \frac{4}{3 + 2\sin\theta}$$

108. Use a graphing utility to graph the polar equation for $e = 1$.

$$r = \frac{4e}{1 - e\cos\theta}$$

Objective 2: Write equations of conics in polar form

109. Which of the following is an equation of the ellipse with vertices at $\left(1, \frac{\pi}{2}\right)$ and $\left(7, \frac{3\pi}{2}\right)$?

(A) $r = \frac{7}{4 - 3\sin\theta}$ (B) $r = \frac{7}{4 + 3\sin\theta}$ (C) $r = \frac{7}{4 - 3\cos\theta}$ (D) $r = \frac{7}{4 + 3\cos\theta}$

110. Which of the following is an equation of the parabola with vertex at $\left(\frac{3}{2}, 0\right)$?

(A) $r = \frac{3}{1 + \sin\theta}$ (B) $r = \frac{3}{1 + \cos\theta}$ (C) $r = \frac{3}{1 - \sin\theta}$ (D) $r = \frac{3}{1 - \cos\theta}$

111. Find the polar equation of a hyperbola with its focus at the pole.
Eccentricity: $e = \frac{7}{3}$; Directrix: $x = -\frac{2}{7}$

112. Find the polar equation of an ellipse with vertices at $(1, 0)$ and $(9, \pi)$.

Objective 3: Use equations of conics in polar form to model real-life problems

113. In a distant galaxy, a planet travels in an elliptical orbit with the planet's sun at one focus. The length of the major axis is $2a$, where $a = 5.17 \times 10^7$ miles, and $e = 0.1945$.
(a) Find the polar equation of the planet's orbit if the focus is at the pole.
(b) Find the minimum distance from the sun to the planet using the equation $r = a(1 - e)$.

(A) (a) $r = \frac{4.97 \times 10^7}{1 - 0.1945\cos\theta}$; (b) 4.16×10^7 miles

(B) (a) $r = \frac{4.16 \times 10^7}{1 - 0.1945\cos\theta}$; (b) 4.16×10^7 miles

(C) (a) $r = \frac{4.16 \times 10^7}{1 - 0.1945\cos\theta}$; (b) 4.97×10^7 miles

(D) (a) $r = \frac{4.97 \times 10^7}{1 - 0.1945\cos\theta}$; (b) 4.97×10^7 miles

114. A comet with a parabolic orbit with the sun as its focus is 710,000 miles from the center of the sun at its closest point.
(a) Find the polar equation of the parabola with its focus at the pole and vertex at $(710{,}000,\ 0)$.
(b) Find the distance between the center of the sun and the comet when $\theta = 75°$.

(A) (a) $\dfrac{1{,}420{,}000}{1-\cos\theta}$; (b) 1,128,000 miles

(B) (a) $\dfrac{1{,}420{,}000}{1+\cos\theta}$; (b) 1,916,000 miles

(C) (a) $\dfrac{1{,}420{,}000}{1+\cos\theta}$; (b) 1,128,000 miles

(D) (a) $\dfrac{1{,}420{,}000}{1-\cos\theta}$; (b) 1,916,000 miles

115. A comet with a hyperbolic orbit has the sun as a focus.
(a) Find the polar equation of the orbit with the focus at the pole, directrix at $x = 1{,}400{,}000$, and an eccentricity of 1.17.
(b) How close does the comet come to the surface of the sun? (Assume all measurements are in miles and the diameter of the sun is 864,000 miles.)

116. A satellite travels in an elliptical orbit with a planet as a focus. It is 153 miles above the surface of the planet at its lowest point, and 109,000 miles above the surface of the planet at its highest point.
(a) Write a polar equation of the orbit with the planet at the pole and one of the vertices at $(112000,\ 0)$. (Assume the planet has a radius of 3000 miles.)

(b) Find the distance between the surface of the planet and the satellite when $\theta = \dfrac{\pi}{3}$.

Answer Key for Chapter 6 Topics in Analytic Geometry

Section 6.1: Lines

Objective 1: Find the inclination of a line

[1] (D)

[2] (B)

[3] 0.62 radians, 35.5°

[4] 1.71 radians, 98.1°

Objective 2: Find the angle between two lines

[5] (A)

[6] (A)

[7] 1.02 radians, 58.3°

[8] 1 radian, 57.5°

Objective 3: Find the distance between a point and a line

[9] (B)

[10] (D)

[11] 5.66

[12] (a) $11\sqrt{2}$; (b) 22

Section 6.2: Introduction to Conics: Parabolas

Objective 1: Recognize a conic as the intersection of a plane and a double-napped cone

[13] (A)

[14] (D)

Answers may vary. Sample answer:

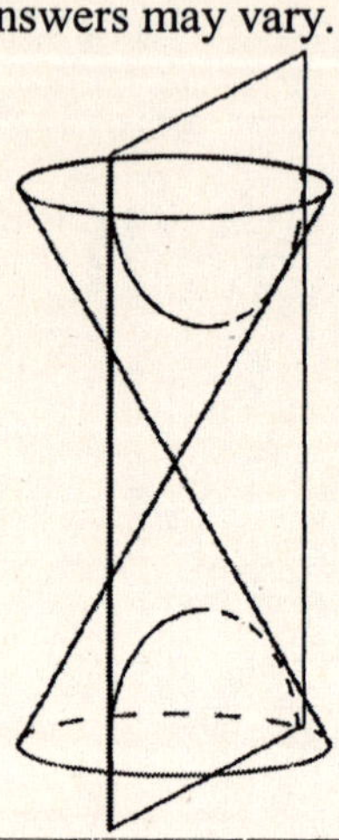

[15]

Answers may vary. Sample answer:

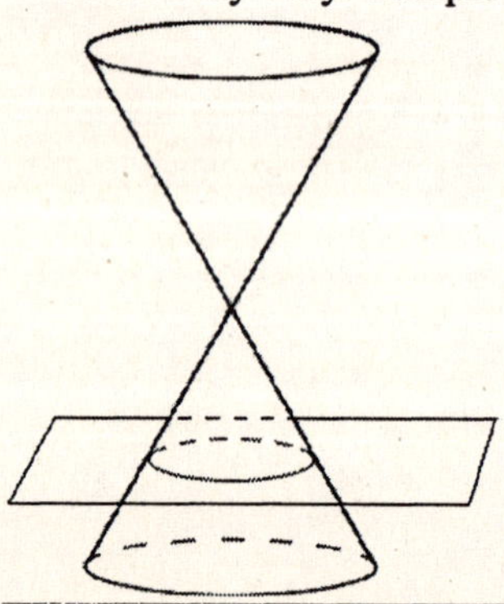

[16]

Objective 2: Write the standard form of the equation of a parabola

[17] (C)

[18] (C)

[19] $(x-1)^2 = \frac{25}{4}(y+4)$

[20] $x^2 = -24y$

Objective 3: Use the reflective property of parabolas to solve real-life problems

[21] (A)

[22] (C)

[23] (a) $y = \frac{108}{22{,}201}x^2 + 2$; (b) 2.175 meters

[24] 1.56 ft

Section 6.3 Ellipses

Objective 1: Write the standard form of the equation of an ellipse

[25] (C)

[26] (C)

[27] $\frac{(x-5)^2}{81} + \frac{(y-2)^2}{16} = 1$

[28] $\frac{x^2}{81} + \frac{y^2}{49} = 1$

Objective 2: Use properties of ellipses to model and solve real-life problems

[29] (B)

[30] (D)

[31] $\dfrac{x^2}{1.23\times 10^{10}}+\dfrac{y^2}{3.48\times 10^9}=1$

[32] $\dfrac{x^2}{1600}+\dfrac{y^2}{441}=1$

Objective 3: Find the eccentricity of an ellipse

[33] (D)

[34] (A)

[35] $\dfrac{\sqrt{11}}{6}$

[36] $\dfrac{\sqrt{55}}{8}$

Section 6.4: Hyperbolas

Objective 1: Write the standard form of the equation of a hyperbola

[37] (D)

[38] (B)

[39] $\dfrac{(x-6)^2}{36}-\dfrac{(y+1)^2}{9}=1$

[40] $\dfrac{x^2}{4}-\dfrac{y^2}{9}=1$

Objective 2: Find the asymptotes of a hyperbola

[41] (C)

[42] (B)

[43] Asymptotes: $y = \pm \frac{5}{3}(x+2)+4$

[44] Asymptotes: $y = \pm \frac{4}{5}x$

Objective 3: Use properties of hyperbolas to solve real-life problems

[45] (A)

[46] (B)

[47] $(-4000,\ 8046.97)$

[48] (a) The ship would reach shore 54 miles from the master station.
(b) 0.000968 second
(c) The ship is at $(174.9,\ 80)$.

Objective 4: Classify a conic from its general equation

[49] (D)

[50] (C)

[51] Parabola

[52] Hyperbola

Section 6.5: Rotation of Conics

Objective 1: Rotate the coordinate axes to eliminate the xy-term in the equation of a conic

[53] (A)

[54] (A)

$$\frac{(x')^2}{4} - \frac{(y')^2}{4} = 1$$

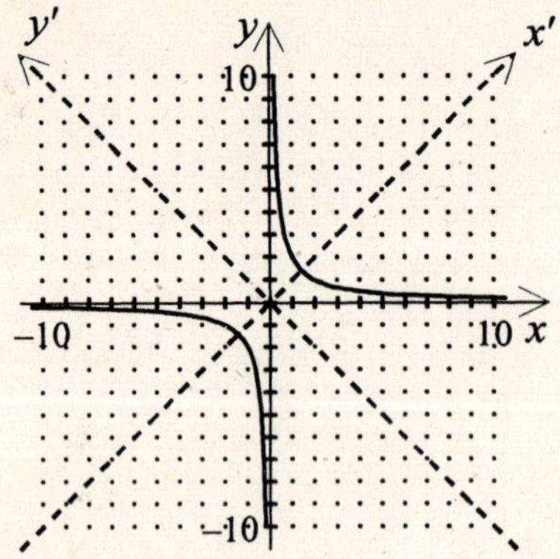

[55]

$\theta = -33.69°$

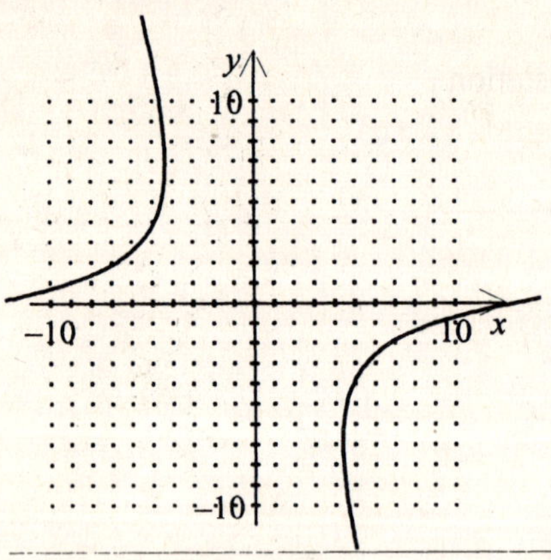

[56]

Objective 2: Use the discriminant to classify a conic

[57] (B)

[58] (D)

ellipse

$$y = \frac{-(-9x+3) \pm \sqrt{(-9x+3)^2 - 4(5)(5x^2+5x)}}{2(5)}$$

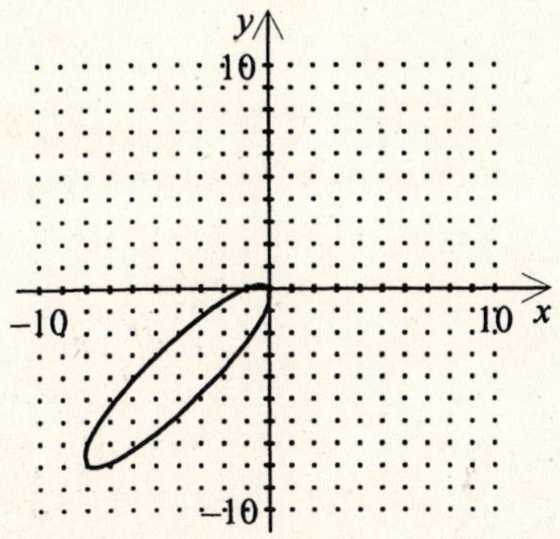

[59]

Ellipse

$$y = \frac{-(2x+5) \pm \sqrt{(2x+5)^2 - 4(5)(5x^2-4x-61)}}{2(5)}$$

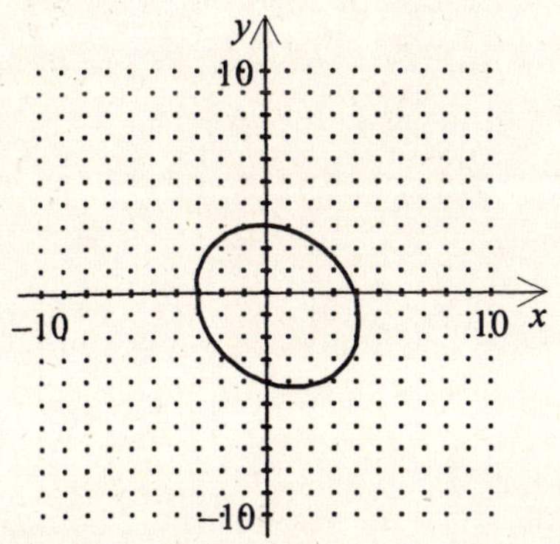

[60]

Section 6.6: Parametric Equations

Objective 1: Evaluate a set of parametric equations for a given value of the parameter

[61] (B)

[62] (C)

[63] $x = -1 + \sqrt{2},\ y = -3 + \sqrt{2}$

[64] $x = 54,\ y = -3.29584$

Objective 2: Sketch the curve that is represented by a set of parametric equations

[65] (C)

[66] (C)

$$\frac{x^2}{4} + \frac{y^2}{25} = 1$$

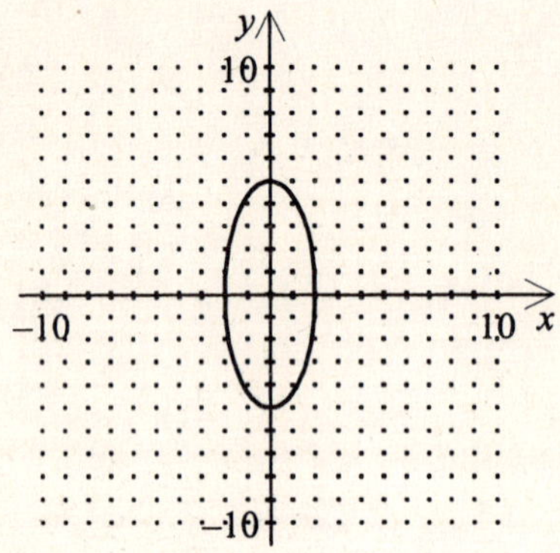

[67]

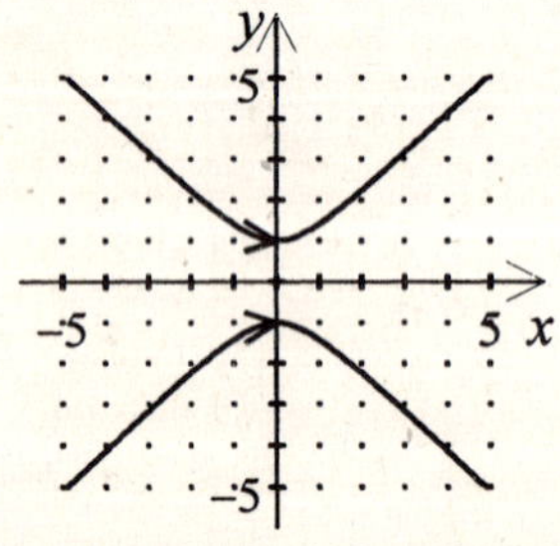

[68]

Objective 3: Rewrite a set of parametric equations as a single rectangular equation

[69] (A)

[70] (D)

[71] $\frac{x^2}{25} - \frac{y^2}{16} = 1$

[72] $x = 4\left(\frac{y+3}{7}\right)^2 + 5$

Objective 4: Find a set of parametric equations for a graph

[73] (A)

[74] (D)

[75] $x = 3t$ and $y = -t^2 - 3t + 10$

[76] $x = t$ and $y = t^3 + t + 2$

Section 6.7: Polar Coordinates

Objective 1: Plot points on the polar coordinate system

[77] (C)

[78] (B)

$\left(4, \frac{11\pi}{6}\right), \left(-4, -\frac{7\pi}{6}\right)$

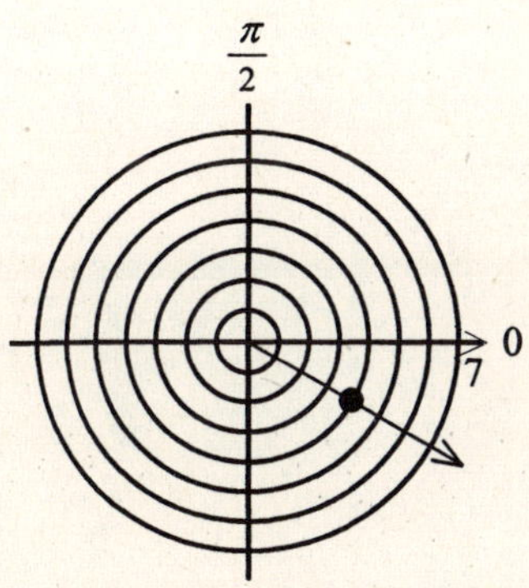

[79]

$(4,\ 5),\ (-4,\ -4.42)$

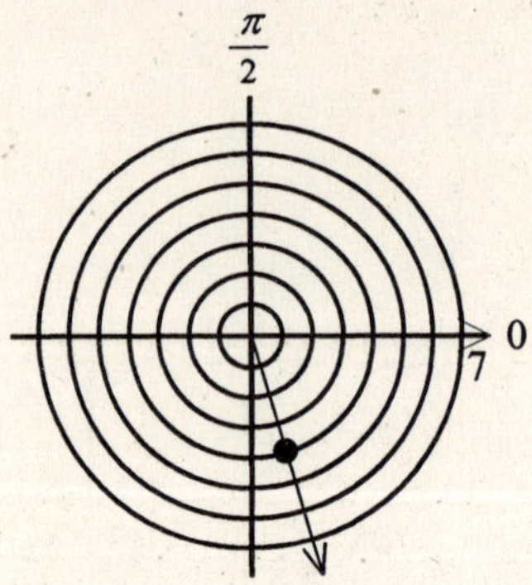

[80]

Objective 2: Convert points from rectangular to polar form and vice versa

[81] (C)

[82] (A)

[83] $\left(4,\ \frac{2\pi}{3}\right)$

[84] $(-9, 0)$

Objective 3: Convert equations from rectangular to polar form and vice versa

[85] (B)

[86] (D)

[87] $r = 4\sin\theta$

[88] $45x^2 + 49y^2 - 12x - 9 = 0$

Section 6.8: Graphs of Polar Equations

Objective 1: Graph polar equations by point plotting

[89] (B)

[90] (C)

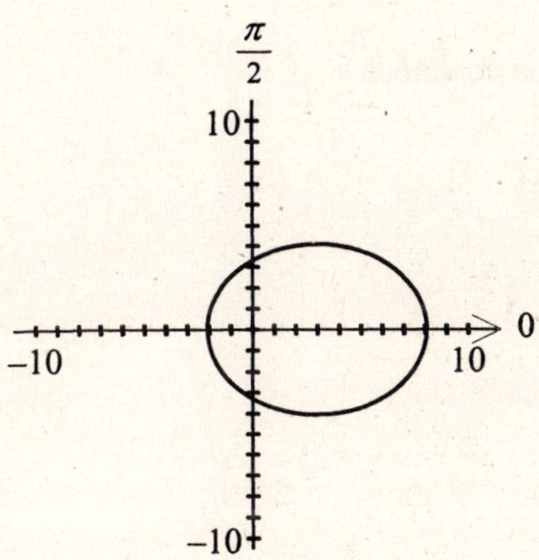

[91] ______________________________

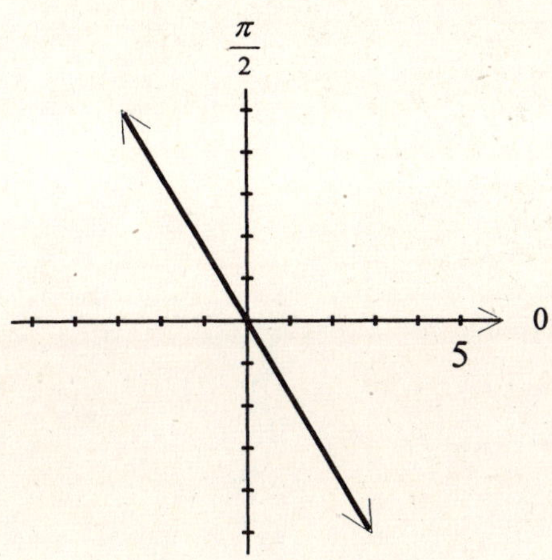

[92] ______________________________

Objective 2: Use symmetry to sketch graphs of polar equations

[93] (D)

[94] (D)

[95] The graph is symmetric with respect to $\theta = \frac{\pi}{2}$.

[96] The line $\theta = \frac{\pi}{2}$: The graph is symmetric with respect to the line $\theta = \frac{\pi}{2}$.
Polar axis: The graph is symmetric with respect to the polar axis.
The pole: The graph is symmetric with respect to the pole.

Objective 3: Use zeros and maximum r-values to sketch graphs of polar equations

[97] (D)

[98] (A)

[99] $8;\ \theta = 0,\ \frac{\pi}{4},\ \frac{\pi}{2},\ \frac{3\pi}{4},\ \pi,\ \frac{5\pi}{4},\ \frac{3\pi}{2},\ \frac{7\pi}{4}$

[100] $6\sqrt{3} + 6;\ \theta = \frac{\pi}{6},\ \frac{5\pi}{6}$

Objective 4: Recognize special polar graphs

[101] (A)

[102] (D)

[103] Circle

[104] Rose

Section 6.9: Polar Equations of Conics

Objective 1: Define conics in terms of eccentricity

[105] (C)

[106] (C)

Ellipse

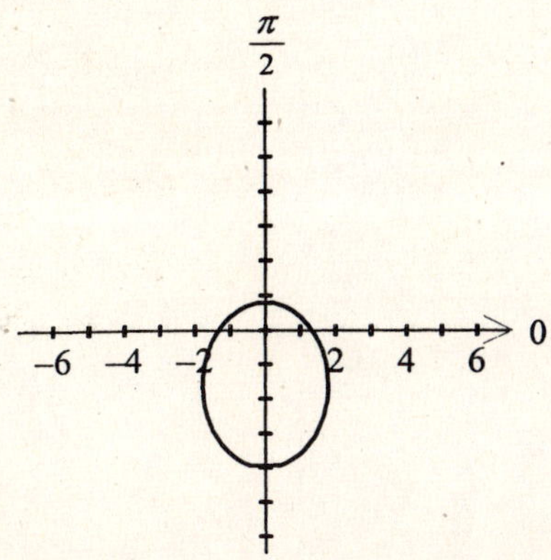

[107] ______

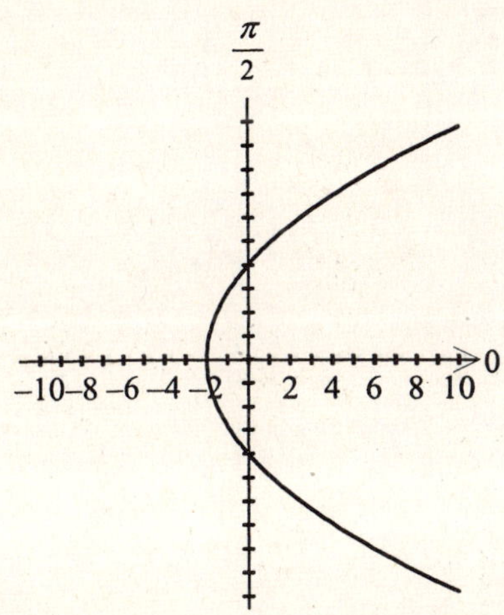

[108] ______

Objective 2: Write equations of conics in polar form

[109] (B)

[110] (B)

[111] $r = \dfrac{2}{3 - 7\cos\theta}$

[112] $r = \dfrac{9}{5 + 4\cos\theta}$

Objective 3: Use equations of conics in polar form to model real-life problems

[113] (A)

[114] (C)

[115] (a) $r = \dfrac{1{,}638{,}000}{1 + 1.17\cos\theta}$; (b) 323,000 miles

[116] (a) $r = \dfrac{6133.34}{1 - 0.945238\cos\theta}$
(b) 8629.81 miles